# WINGS OF FAITH

A VIETNAM VETERAN'S JOURNEY THROUGH
WAR, SURVIVAL, AND GOD'S CALLING

## COL. DAVID O. SCHEIDING, USAF (RET)

978-1-965552-63-6 (Paperback)
978-1-965552-64-3 (Hardback)

*Library of Congress Control Number:* 2026901668

BOOKWRIGHTS HOUSE

admin@bookwrightshouse.com
☎ (213) 286 6700

COL. DAVID O. SCHEIDING, USAF (RET)
is also the Author of

"Hank, An Angel Dog"
"The Long Return"
and
"Don't Fly Today"

# DEDICATION

I DEDICATE THIS BOOK to my wife of sixty-three years plus. God provided me with the absolute best wife I could ever have hoped to find. I know it was God's plan for our paths to cross and for us to fall in love as part of his overall plan for both of us. God has a plan for all of us if we listen and believe in him and his son Jesus. I know God saved my life at least three times in Vietnam to ensure that I would be here now to take care of my wife as she struggles with dementia.

# ACKNOWLEDGEMENTS

I WOULD LIKE TO thank Mr. Mike Wolf for encouraging me to write this memoir of my life. I am truly humbled to think that someone would actually think my life has been significant enough to be documented. If so, I thank God for all of his blessings and guidance throughout my entire life. I was following his will and guidance to accomplish what he needed me to do as part of his overall plan for his creation. He gave me what I needed, not always what I wanted. I consider my life to be a testament to God and he deserves all of the honor and glory. I am merely the vessel he used for his plan.

# PREFACE

I AM NOW EIGHTY-THREE years old, retired, and I now have a lot of time to reflect back over the path that I have traveled throughout the course of my life. As I look back over my life, and from all of the empirical evidence available, I can only conclude that God does exist, and he does have a plan for each and every one of us as part of his overall plan for his creation.

I reached this conclusion based on all of the empirical evidence present over the past eighty-three years of my life. I realized that a higher power had to have been guiding me in order to explain everything that has happened to me over the course of my life. It could not have been coincidence or random events that occurred which resulted in the obviously defined path on which I have traveled.

A person cannot look at individual events, decisions, and actions or singular random events that occur in the course of one's life. We have to take into account the totality of the sum of all these events, decisions, and actions in order to see how they all fit together over the entire span of one's life. When we do, we can see that God has a specific plan and purpose for each of us. The following pages document the empirical evidence of all those individual events, decisions, and actions that have occurred in my life, which produced the plan that God had for me.

Wings of Faith: A Vietnam Veteran's Journey Through War, Survival, and God's Calling.

By Col. David O. Scheiding, USAF (Ret.)

# TABLE OF CONTENTS

# CHAPTER 1

# The Soil of Iowa

My life began as a blessing from God. We do not get to choose our parents, but God does. I do firmly believe that God chooses our parents as part of his overall plan for each member of his creation. As part of his plan for each of us, I believe that he guides each individual to meet others that he has chosen, who after that fall in love and get married. This creates a family of a mother and father, which fulfills his guidance in the Bible to go forth, be fruitful, and multiply. This was evidenced when God directed both Adam and Eve in the beginning and after that Noah and his family after the flood, to be fruitful and multiply. This was required to first populate the earth and after that to replenish it after God destroyed it. This tells me that the sanctity of marriage, as being one man and one woman, is quite important to God. He destroyed the earth and all of his creation due to wickedness of mankind during Noah's time. I was blessed and quite fortunate in that God chose my mother and father to be my parents. One's birth is when God's plan begins for each of us.

I was born November 14, 1942, into a world of hard work, faith, and midwestern grit. Marshalltown, Iowa, was not a place of glamour or wealth, but a small town built on community, industry, and farming families who understood sacrifice long before the word became a slogan. The men worked long hours in foundries and fields; the women kept homes running through sheer willpower and resourcefulness. Children, like me, learned early that nothing came easy, but everything worth having could be earned through honesty, humility, and discipline.

My father's parents immigrated to the United States in the late 1890's. My grandfather was from Germany, while my grandmother was from Bohemia. My grandfather worked as a railroad train engineer, and I did not see him much. My grandmother had the most influence on me as a child. Nevertheless, both my grandfather and grandmother were extremely kind and giving. In fact, before I was born, they adopted a 5-year-old little girl that lived up the street from them. Her parents could no longer take care of her and asked my grandparents if they could adopt her. Without hesitation, they both said yes. This resulted in my father now having a quite young little sister. My grandmother was perfect for this, because she always said, she would have liked to have run an orphanage. It was also good for me because this little girl was more like a big sister to me growing up instead of a very young aunt.

My earliest memories are of my grandmother's house—a small home with a chicken coop and small garden in the back yard where she raised chickens and vegetables for food. I was in the back yard one day when one of the roosters attacked me. My grandmother immediately chased down that rooster and we had him for dinner that night. Both of my parents worked, so my grandmother became my daytime caretaker, her gentle guidance shaping much of my early character. She was the kindest and gentlest person I have ever known and whose voice was soft yet unwavering. It was in her home that I first learned about prayer, about gratitude, and about a God who watched over every small detail of life. Her faith was not dramatic; it was quiet and steadfast, woven into everything she did.

I was blessed to grow up in a family where my father's integrity and my mother's resilience balanced each other perfectly. My father was a kind, honest man steady as a metronome. He worked long hours at the foundry, a dangerous and exhausting job where molten metal molds would occasionally explode, burning him and spraying sparks and heat across his clothes. But he never complained. Instead, he saw his work as a means to provide for his family, and that responsibility was something he bore with pride.

My mother's childhood had been much harsher. She and her brother were abandoned at the Marshalltown courthouse as a young girl, left by her own mother in an act of cruelty that haunted her for years. She was raised in a broken family marked by alcoholism and poverty; she learned survival skills early. Yet, she also found faith early, choosing to cling to God's promises when her earthly family failed her. My mother and father actually met at church. When my father asked her to marry him, she kept saying let's don't destroy our relationship by getting married. She thought all marriages would replicate the dysfunction and fighting she had seen in her own family. It took my father three years with persistent, patient love to convince her otherwise. Her story shaped much of who I became, because I watched her overcome bitterness with grace and hardship with courage.

I learned two quite important lessons from my parent's families growing up a youth. The first lesson was how my mother was treated by her mother. The second lesson was on my dad's side of the family when I learned how the devil uses greed to impact a family as he works his evil. When my dad's father, my grandfather, passed away, one of his birth sisters and her husband were tempted by the devil with greed. Their greed manifested itself when they were able to take control over a major portion of my grandmother's inheritance after the death of her husband.

My grandfather owned a large farm, on which dad's other birth sister and her husband were living. They were paying rent to my grandfather and grandmother until his death. The rent from this farm would have taken quite good care of my grandmother's needs as far as income after my grandfather's death. Because of the greed of this sister and her husband, they were causing significant disruption in the family as they pursued their goal to gain control of the farm. This disruption was pitting brothers against sisters and literally tearing the family apart. This was hurting my grandmother so much that she finally asked my dad, his brother, and the other two sisters to sign the farm over to the greedy sister and her husband. She did not want to fight the legal battle due to the discord that it was causing within the family. My dad, his brother, and the two sisters agreed to do so in order to keep the family together and to honor their mother's wishes. This showed me how kind and gentle my grandmother was and why she would adopt another child so late in life.

The loss of the farm income to my grandmother caused her to have to take in the laundry of others in order to make ends meet. She also had to plant an additional garden on a piece of land that was part of her home to help with food. I do not know how anyone could treat their mother that way, especially when they claimed to be quite religious members of a church. This had to be Satan at work as he battled God.

Our home was simple but full of love. We weren't wealthy by any means, but God provided what we needed. In fact, my parents did not even have a car when I was born. My parents taught me that character was more important than possessions. I watched them make sacrifices for me without complaint. I also watched how deeply they respected God's word, not only on Sundays but every day of the week. My mother read her Bible faithfully, and my father modeled humility in everything he did. Those lessons formed the bedrock of my life, even in moments when fear and doubt threatened to shake me years later.

My parents raised me to believe in God and his son Jesus as well as set a good example for me. This is why I feel that I was truly blessed to have them as my parents. All the empirical evidence supports the fact that God had to be guiding my life from birth. This to me also supports the fact that life is not a series of random events or incidents that happen but are all pieces of God's plan for each of us. When all these events are placed together, they form the path that he has chosen for each of us. It also reflects how God's plans for all of us mesh together for his overall plan for his total creation.

# CHAPTER 2

# Childhood Lessons

GROWING UP IN IOWA in the 1940s and 50s was a different world from today. We rode bicycles without helmets, played outside until dusk, and rarely locked our doors. Marshalltown was a close-knit community, a place where neighbors looked out for each other. But it was also a place where children were taught to work early. By the time I was a teenager, I had held various jobs from having a morning paper route, mowing lawns, to helping at my father's grocery store and meat market. My father had decided he wanted to move from the quite harsh foundry environment to a less dangerous work environment. He and my mother were able to purchase a small neighborhood grocery store and meat market. He learned how to be a butcher and when I asked him if he would teach me how to butcher, he was quite happy to do so. Our store was open from 6:00 a.m. to 8:00 p.m., ever day except Sunday. That was God's Day.

Since I was able to assist my father with the butchering duties, we only had one other employee to help my mother with cashier duties and restocking shelves These were quite long days for both my mother and father. When a customer would want to purchase a side of beef from the meat market, we had to cut and wrap the side of beef after 8:00 p.m. when the store was closed. I could tell that as our business increased on the meat market side, it was beginning to take a toll on my father. I saw an opportunity to be able to help him, so I offered to do the cutting and wrapping of the sides of beef by myself. He liked the idea and said that if I did that, I could keep all the profit from the sale of the sides of beef. I could cut and wrap a side of beef in two hours. So, as a teenager, I became an entrepreneur. I made enough profit that I was able to purchase a used red 1955 Ford convertible. This car became a quite important element later in my life during my sophomore year in college. As far as the store, I was able to run the store for my parents when they wanted to take a vacation. I also was able to fill in for my father full time when he cut his finger off on a band saw. I did this until he was able to return to work. Those experiences taught me responsibility and self-reliance. They also taught me about people—their kindness, their flaws, and their struggles.

The store was more than a business; it was a gathering place where locals came not to shop but to talk. I would watch my father cut meat for customers while listening to conversations about farming, weather, and politics. He treated every person who came through the door with respect, whether they were wealthy or barely scraping by. That attitude of fairness stayed with me throughout my career. Leadership, I learned, wasn't about rank or status—it was about how you treated the people around you.

But while my father was shaping my work ethic, my mother was shaping my faith. Her childhood trauma could have turned her bitter, but instead, it drove her closer to God. She was determined to give me a stable, loving home—something she had been denied. I watched her forgive family members who had wronged her deeply, even her own mother. She often quoted Exodus 20:12 "Honor thy father and thy mother: that thy days may be long upon the earth." She taught me to have faith in God as he has a plan for everyone, and we need to listen and obey. This came home to me later in Vietnam when survival often came down to a split-second decision—or a voice I couldn't explain telling me don't fly today.

As far as sports go as a youth, I did play basketball and tennis while in high school. God gave me the talent to play tennis, and I essentially taught myself how to play. Again, as I reflect back over my life, this too is further empirical evidence that God was guiding my life, and this was part of his plan for me. Tennis was an important factor during my college years. This was another piece of my life's puzzle on my path through life. God and my parents prepared me well for my life after high school.

# CHAPTER 3

# The Call of the Sky

WHILE MY FAMILY GROUNDED me in faith and values, my dreams were already airborne. I was fascinated by airplanes from a young age. Marshalltown's open skies became my playground, and I would often lie in the grass, watching contrails streak across the horizon. Airplanes seemed like freedom incarnate — machines that defied gravity and fearlessly soared above the earth. Even today, I still look up when F-16s from the Texas National Guard fly over my house here in San Antonio as it reminds me of my days flying the F-111. I also watch as the C-5s from the Air Force Reserve Wing fly overhead, because it reminds me of a significant experience that greatly impacted my life when I was the lead structural engineer for the C-5 fleet.

In high school, I read books about aviation. Charles Lindbergh, Chuck Yeager, and the World War II fighter aces became my heroes. Their courage inspired me, but I didn't want to read about their adventures — I wanted to live them. Aviation wasn't simply a career goal; it was a calling. I chose Iowa State University (ISU) to study Aerospace Engineering, because at that time, ISU was in the top 10 engineering universities in the country. I wanted to design airplanes.

By the time I entered ISU, my path was becoming clear. ROTC was mandatory for the first two years at state universities, and I discovered that I enjoyed the structure and discipline. By my junior year, I had the opportunity to sign up for Advanced

ROTC. I decided I would signup if I could pass the Air Force flight physical. Passing that exam felt like the first door swinging open to my destiny. When I graduated with a degree in Aerospace Engineering, I was also commissioned as a second lieutenant in the United States Air Force. The boy who once watched contrails from his backyard was about to make his own.

It was during my sophomore year at ISU when that red 1955 Ford convertible became an important factor to me. The Greek fraternity system was quite strong at ISU, and I had joined the Delta Tau Delta fraternity. It was during this year and one of my fraternity brothers had been dating a young lady that I considered to be the most beautiful women I had ever seen. When he broke up with her, I asked him if he cared if I asked her for a coffee date. He said he had no problem with that, since he had moved on. When I got up enough nerve to call her, she of course did not know me from Adam. I told her that I was the Delt that owned the red 1955 Ford convertible that she had seen in the fraternity house parking lot. With that she agreed to have a coffee date with me at the Student Union.

After our coffee date, our first real date was to a Kingston Trio music concert at ISU. I was quite shy in high school around girls, but I was never shy around Jan. From that date on, we would go to the library each night to study, and we dated each weekend thereafter. God had made it possible for us to meet as part of his plan for both of us.

I later found out, as additional empirical evidence, that God had to have been guiding us to meet because Jan had wanted to go to college at the University of Iowa and not ISU. The University of Iowa is located in Iowa City, which is approximately one hundred miles to the east of Ames, Iowa, where ISU is located. Jan's home was on a farm near the small town of Underwood, Iowa. Underwood is located on the western border of the state of Iowa. Jan's parents thought that the University of Iowa, being an additional one hundred miles farther away, was too far away from home for their daughter to go to school. Jan's parents convinced her to go to ISU instead of the University of Iowa. There is no doubt in my mind that God had to be at work in the decision as to where

Jan was going to school. God was guiding her path in order for us to meet as part of his plan for me.

During the summer between our freshman and sophomore years, we dated. Each weekend I would drive from Marshalltown to her parents' farm that was located near Underwood, Iowa. Underwood was approximately 180 miles from Marshalltown. We did not let that stop us from seeing each other. I put a lot of miles on that 1955 Ford convertible that summer. This was the time when my parents owned the neighborhood grocery store and meat market.

During the week, I would butcher and help run the store, however, on the weekends, I had to go see Jan. My parents were quite good about me going to see Jan on the weekends. In fact, my mom told my dad that she thought that Jan was the one. My mom and God did know that Jan was the one for me. At the end of that summer, I proposed to Jan. Fortunately for me, she said yes!—which I know now was all part of God's plan.

I had done well in the fraternity and had been elected as vice president at the end of our freshman year. As I mentioned earlier, I had taught myself how to play tennis while growing up and had played in high school. I decided to try out for the ISU tennis team and fortunately was successful there too. Here again is where I know that this was all part of God's plan for me. I never had a tennis lesson in my life, but by my senior year at ISU, I was the number one player on the ISU tennis team. This too became a significant factor in Jan's and my life that helped us when we needed it.

Since I had proposed to Jan during the summer between our freshman and sophomore years, I was more interested in Jan, school, and tennis as opposed to the fraternity. Jan seemed to feel the same way, so we decided to get married between the fall and winter quarter break of our sophomore year. This was quite a surprise to both of our parents, as well as to my fraternity and Jan's sorority. My fraternity housemother did not approve and told me that I was way too young to get married. I did not truly care what she thought as long as Jan and my parents approved.

I told my mom first about our plan. She said, "Go tell your dad." My dad did not say much but did ask me if we planned to finish school. Both Jan and I had every intention of finishing our degrees.

Neither of us had any plans to quit school. With this understanding, Dad agreed to continue to pay for my college tuition.

After getting my parents' approval, the next step was to determine how Jan's parents felt about her getting married. I felt I had a good chance with her parents since I knew they had gotten married when her mother was quite young. I first asked her dad if I could marry his daughter. Herman (Jan's dad) was a farmer and also worked for the Union Pacific Railroad. Herman was a quite wise man who, like my mother, had to quit school when he was young in order to help support his family. Herman was one of twelve children in his family. He was a quiet and gentle man. When I asked him if I could marry his daughter, he listened politely in his quiet manner. After I told him how I felt about his daughter, I finished up with my promise to him that I would always take quite good care of her. He smiled at me and shook his head in an affirmative manner.

When we told Jan's mother about our plans, her only concern was that Jan would finish school and get her degree. When we both confirmed that we were determined to finish college, she also agreed to our request for her approval. In addition, Jan's parents also agreed to continue paying for Jan's college tuition, which was great news to me.

With our parents agreeing to continue to pay for our college tuition, we next had to figure out how to support ourselves. This is where I know that God's plan for my parents to go into the grocery store business entered in as help for Jan and me. In addition, to continue paying for my college tuition, they also volunteered to allow us to get our food from the store. Since ISU was located only thirty miles from Marshalltown, it was no problem for us to drive to Marshalltown every couple of weeks to get food. My parents also agreed to pay our rent since it was about the same as my monthly fraternity dues. I know that God had laid all the groundwork for us to get married as part of his plan for us.

With the approval of both our parents, Jan and I were married on November 25, 1962, between fall and winter quarters of our sophomore year. At that time, ISU was on a quarter system and not a semester system. Each college year consisted of three quarters instead of two semesters.

As I now think back about how our parents must have thought about us getting married so young, it could only have been God's guiding hand making everything possible. Growing up, both of our parents believed in God and raised both Jan and me to also believe in God. I will say that at that age, I was not thinking about all that was happening in my life, that it was actually all part of God's plan for me. I think that people at that age are more concerned with living their lives as opposed to thinking about God even having a plan for each of us.

Now, after sixty-three years of marriage, it is much easier to reflect back over the years to see how God had to have been guiding our lives for all of it to happen. We had to listen to him and trust him. I also now wonder what my fraternity housemother would now think about Jan and me getting married at such a young age.

After we got married, Jan moved out of her sorority house, and I moved out of my fraternity. We rented an apartment and began our lives together. We were quite happy as we continued on with our education. God blessed us during this time by having our parents pay our tuition and my parents providing our food and paying our rent. God also provided us with other opportunities that helped us support ourselves. I was selected to become a student manager for the ISU basketball team. My tennis coach, who was also an assistant basketball coach, helped me get this position. As payment, I was given student food vouchers to eat at the Student Union. This helped us significantly during basketball season with our food supplies. I also was given free basketball tickets for my parents to come over to ISU for the home basketball games. They truly enjoyed the basketball games. It was also a small way for me to pay back my parents for all of their help.

I also got a job during the summers between our sophomore, junior, and senior years to work for the University's Landscape Department. This allowed us to go to the Dairy Queen once a week to get some ice cream as a treat for ourselves. It was great, and we were extremely happy. This helped us a lot financially during this time period. There were two other financial factors that also helped us, which I have to attribute back to God's plan for us. This first was the decision for me to apply for Advanced ROTC. The

Air Force paid a small stipend to each Advanced ROTC student each quarter. We would get a check each quarter from the Air Force. This truly helped us during our junior and senior years at ISU. It always seemed to arrive at the time when we truly needed it. I remember one time when we did not have $0.19 to buy a loaf of bread. The next day, the Air Force check arrived. God knew our needs and took care of us.

Another financial factor that God provided us occurred during our senior year. My tennis coach had put me on a tennis scholarship for that year. I know that it had to be God's work because ISU did not offer tennis scholarships. This is more empirical evidence that God does exist and that he has a plan for all of us.

Another piece of empirical evidence that God was guiding me became apparent while playing tennis during my senior year. Since God had led my tennis coach to provide me with a tennis scholarship at a school that did not previously have tennis scholarships, it was quite important for me to remain the number one player on the team at ISU. It was during my senior year that a quite good tennis player was also attending ISU. He had played number one for the freshman team during his freshman year as I had done. At that time, freshmen athletes were not allowed to play on the varsity teams, and we had to wait until our sophomore year to play varsity sports. I had played number one during my freshman year, and after that I was able to play at the number three position during my sophomore and junior years. I was number three behind the number one player who had been the Nebraska state champ while he was in high school. Our number two player had been the South Dakota state champ while he was in high school. I was not the Iowa state champ, but I was part of the Iowa State High School District doubles team champions.

As a senior on the tennis team, I was now playing the number one position as the previous number one and two players had graduated. I had to earn that position at the start of the fall tennis season when the tennis coach held a challenge tournament to determine what position each of the members of the team would play before the start of the season. During this tournament I was able to defeat the now sophomore tennis player that had played number one as a freshman. He did, nevertheless, obtain the number two position on the team.

This individual's name was Bill O'Brian, and I remember him well because he was a quite good tennis player. I do not remember the names of any of the other members of our team my senior year, but I do remember him. I know that God had helped me defeat him and secure the number one position, which led the coach to put me on scholarship.

Once the spring tennis season started, other team members could challenge for a higher position on the team. Bill O'Brian was a baseline tennis player and was extremely adept at keeping the ball in play, which would allow opponents to make mistakes and essentially beat themselves. My game was more of a serve-and-volley game, which meant that I would attempt to overpower my opponent. I would serve and after that make my way up to the net in order to put away the opponent's return. To defeat a baseline player, a serve-and-volley player truly has to be on his game in order to defeat someone who is doing everything he can to keep the ball in play until a mistake is made. Approximately midway through the season, Bill challenged me for my position. I knew that it was going to be quite difficult for me to defeat him again. He had been playing quite well.

This is when I now know that God stepped in to help me maintain my number one position and be able to retain my scholarship. If I had lost, it would have been difficult for my coach to defend me having a scholarship while Bill O'Brian did not. This would have been especially true since ISU supposedly did not offer tennis scholarships. On the day that the challenge match was scheduled to be played, it was raining. The outdoor courts at ISU had lay-cold surfaces. This is a green coating that is placed over a hard surface such as concrete and provides an excellent surface to play on. It plays faster than a clay court surface but slower than a concrete surface. If a serve-and-volley player is any way off of his game on a fast surface, baseline players can be quite effective. Since it was raining, the coach had to move the match to our indoor court, which was in the armory building at ISU. The armory is where ISU played basketball. The basketball court was placed on the concrete floor of the armory. During the spring quarter, the basketball floor is removed, and after that, became our indoor tennis court.

The floor of the armory consisted of a polished concrete surface slab. This type of surface is a much better surface for a serve-and-volley game because it plays so fast. Slower surfaces give the advantage to a baseline player. Having watched Bill play during the first half of the season, I was not sure that I could beat him again. He had truly improved from our first match in the fall to the now spring quarter. This is where I feel that God intervened with the rain because it forced us to play indoors on a surface that was more suitable to my type of game. The polished concrete floor of the armory was much faster than the lay-cold surfaces on the outdoor courts. I again was able to defeat Bill relatively convincingly enough that he never challenged me again. I kept my number one position and did not put my coach in a position of having to defend my scholarship at a school that did not previously have tennis scholarships. This had to be God's plan for me.

As I reflect back over all the circumstances that had to align for Jan and I to meet, it became obvious to me that these factors were all part of the empirical evidence that God was guiding my life. Let's review some of those factors: ISU was the only college I considered attending. I found out later Jan had wanted to go to Iowa University instead of ISU. Her parents thought that was too far away and insisted she attend ISU. I pledged Delta Tau Delta fraternity, and Jan had been dating my fraternity brother when I first saw her. I had a red 1955 Ford convertible.

When I consider all the intricate interactions that had to occur for all the above to work out, only a higher power could be responsible for all these things to occur. This to me is also further evidence that God does exist, and he has a plan for all of us. For example, as I previously mentioned, ISU was the only college that I even considered attending. Before I left Marshalltown to attend ISU, I did not have any intention of joining a fraternity. I did not even have enough information about the Greek system and fraternities to even have an opinion about them. God, Nevertheless, did have a plan for me.

As far as getting my red 1955 Ford convertible, God had a hand in this also. I had always wanted one as a kid, but I never thought I would have enough money to purchase one. God's plan for my dad when he changed careers and bought the grocery store and meat market gave me the opportunity to learn how to butcher meat. This resulted in me being able to buy that red convertible during my junior year in high school.

Both Jan and I finished our degrees. She graduated with a BS in English, and I graduated with a BS in Aerospace Engineering. Upon graduation and commissioning as a second lieutenant in the Air Force in December 1964, both of us were ready to face life after college.

# CHAPTER 4

# Entering the Air Force

THE AIR FORCE SENT us to Reese Air Force Base in Lubbock, Texas, for Undergraduate Pilot Training (UPT). The year was 1965, and the Vietnam War was beginning to cast a long shadow over military life. For me, though, that year of training was exhilarating. We trained in the T-37 Tweet and the supersonic T-38 Talon, which had replaced the aging T-33, learning not how to fly but how to think, react, and make life-or-death decisions in seconds.

Flight training was rigorous, both physically and mentally. Every student knew that mistakes could be fatal. We were learning to operate machines that demanded precision, and there was no room for error. Some classmates washed out; others pushed themselves to the edge of exhaustion. But I loved every second of it. The first time I soloed, I felt an overwhelming sense of freedom. The world spread out beneath me, vast and inviting, and I knew I had chosen the right path.

The first day at Reese Air Force Base was not unlike the first day of boot camp, except instead of rifles and barracks, we were handed flight manuals and helmets. The instructors didn't smile much. They didn't need to. Their presence alone told us that we had stepped into a world where mistakes were not tolerated. "Gentlemen," one of them said on that first morning, "airplanes are not forgiving. If you want to survive, learn quickly. If you want to thrive, listen even faster."

We began with ground school, long hours of aerodynamics, aircraft systems, and regulations. For a small-town kid from Iowa who had studied engineering, much of it fascinated me. Lift, drag, thrust, weight—I had read about them in textbooks, but here, they weren't theories. They were life and death.

The first time I climbed into a T-37 "Tweet," my heart pounded so hard I thought the instructor might hear it over the intercom. The Tweet wasn't much to look at—small, stubby, noisy. But to me, it was freedom. When the engine roared to life and we taxied to the runway, I realized I was stepping into a new identity. The wheels left the ground, the earth fell away, and I was flying.

The training was relentless. Each sortie brought new challenges: stalls, spins, precision landings. We learned how to recover from near disasters and how to prevent them in the first place. The T-37 was forgiving enough to let us make mistakes but unforgiving enough to punish carelessness.

By the time I transitioned to the T-38 Talon, the sleek supersonic trainer that looked like a scaled-down jet fighter, I was no longer the nervous small-town kid gripping the stick too tightly. I had become a pilot. The Talon taught us speed, discipline, and the razor-thin margin between control and chaos. Flying at Mach 1 was thrilling, but it was also humbling. One lapse in concentration at that speed, and you were a memory.

Graduation day was one of the proudest moments of my life. My silver wings were pinned to my chest, and I knew I had joined a brotherhood that stretched back to the Wright brothers and forward into the unknown skies of the Vietnam War.

After earning my wings, I had graduated high enough in my class to be able to get my choice of assignments. I decided to request to become an instructor pilot instead of choosing a fighter assignment. The F-4C aircraft was new to the Air Force but the Air Force was assigning new pilots to the back seat and after that they were on their way to Vietnam. Later, the Air Force assigned weapon system officers (WSOs) to the back seats instead of pilots. I knew as an instructor pilot, I would be able to get a lot of flying time and gain significant experience. I wanted a T-38 but received a T-37. I later realized that God was guiding my life with this assignment to T-37s. He gave me what he needed for his plan, not what I wanted. Being a T-37 instructor pilot provided me with extremely valuable experience that I would have never received in the T-38 and greatly shaped my Air Force career as well as saved my life.

Another quite important event also occurred during our time in UPT. Our first son, Douglas Scot Scheiding, was born on February 15, 1966. We were now a family of three.

# CHAPTER 5

# Instructor at Laughlin

THE AIR FORCE DECIDED my first assignment as a T-37 Instructor Pilot (IP) would be at Laughlin Air Force Base in Del Rio, Texas. It wasn't the combat assignment many of us craved, but it was critical work. My job was to take young men, many barely out of college, and teach them to fly the Air Force way.

The T-37 was noisy, temperamental, and demanding. Students often joked that the J-69 engines converted JP4 fuel to noise. But it was also honest. If you treated it well, it would fly beautifully. If you mishandled it, it would punish you instantly.

We arrived at Laughlin AFB in May 1966, after I completed Flight Instructor Training (FIP) at Randolph AFB in San Antonio, Texas, which lasted two months. I began my IP duties after an orientation flight with a stan/eval pilot at Laughlin. I instructed a total of five classes before God provided me with an opportunity that allowed me to advance as an IP. In each class, I was responsible for four or five students. The opportunity that God provided me was a significant challenge, but it turned out to be quite helpful in my career. It was all part of God's plan for me.

The opportunity and challenge that God provided me was being assigned a foreign student named Ammand, who was from Iran to train how to fly. At that time, our country still had good relations with Iran and the Shah of Iran. My student Ammand was also royalty and was number three in line to the Shah. When he was assigned to me, the Squadron Commander called me to his office and told me that I had to get him through the T-37 program. He said that if I did not, Ammand would be put to death on his return to Iran. He had to finish the T-37 program. He indicated that if I could get him through the T-37 program and he washed out of the T-38 program, he would become a C-130 copilot in Iran as he completed his time in the military. No pressure here!

Ammand was a quite nice individual, but he had zero—and I mean zero—flying abilities. He was a real challenge for me due to his total lack of flying capabilities. I knew that I was in deep trouble with him as a student after our first flight. I had never or since flown with someone who lacked totally on coordination and did not have any natural instincts for flying. I knew that it was going to be a real challenge to teach him how to land the T-37.

It normally only required ten to twelve flying hours to teach a new student without any flying experience to fly solo. When we reached twenty hours of flying time, Ammand was nowhere near being able to solo. I informed my Squadron Commander about his total lack of aptitude for flying. He said he understood, but we did not have a choice. The Squadron Commander said, "Let's give him thirty hours and after that see where he is, and after that we will talk again."

When we soloed students in the T-37, we would start out as a dual ride with student and instructor. The student was required to complete three satisfactory landings and a go-around. We would after that complete a full-stop landing and then taxi back to the takeoff end of the runway. Next, we would shut down the right engine, and the instructor would get out of the aircraft. The instructor would then monitor the restart of the right engine. After the restarting of the engine, the instructor would then walk over to the runway supervisor unit (RSU) and monitor the student as he performed three more landings solo. An experienced instructor pilot always manned the RSU and had the responsibility to control the traffic pattern for that runway. It was his responsibility to verify that each aircraft was properly configured for landing and to direct aircraft to go-around if any safety condition was in question. The RSU duty was assigned to different flights of the squadron each day, so you only knew who the RSU supervisor on the days was when your flight had the duty. On the day Ammand was going to reach thirty hours, my flight did have the RSU duty, and everyone in my flight knew about him as a student. On this day, my assistant flight commander was the one assigned RSU duty for the day. His name was Bill Russell, and he was highly experienced. Before Bill left the squadron for the RSU, I informed him that I was going to try to solo Ammand. He looked at me, raised his eyebrows, and said, "Okay, I guess we both will be playing 'you bet your wings' today." He was right!

Ammand and I took off, and we stayed in the traffic pattern in order for me to monitor his three landings and a go-around. On his first landing, I had to assist him. It was not looking good for him. On his second landing, he improved and was able to land the aircraft on his own. It was not a good landing, but I did not have to help him. His third landing was again a little better, but still not a good landing. I decided that we would complete two more landings before I would decide if I was going to solo him. The fourth and fifth landings were marginally acceptable. On the fifth landing, Ammand had called for a full-stop landing. He knew that we were getting low on fuel, and if he was going to be able to complete his three landings, we would have to full stop on this landing. Nevertheless, I still had not made up my mind if I was going to solo him. My decision was made for me when we taxied past our parking row, and we were now headed back to the takeoff end of the runway.

At the end of the runway, I had him shut down the right engine, and I got out of the aircraft. Ammand restarted the engine and was now going to solo. I walked over to the RSU where Bill was performing RSU duties for the day. As I entered the RSU, we both looked at each other, and neither of us said a word. As he took off, all we could now do was sit, wait, and hope for the best.

As we watched Ammand begin his first attempt at landing solo, we could tell that he was fighting the aircraft as he began the final turn. We both watched him quite closely as he struggled around the final turn. He did at least have the aircraft properly configured for landing with the gear and flaps extended. As he arrived over the threshold of the runway, he was higher than he should have been, but he did not elect to go-around. Instead, he pulled power, and both Bill and myself came out of our chairs. Bill had raised the radio that was in his hand to direct him to go-around. Nevertheless, with a T-37, once you pull power it takes approximately twelve to thirteen seconds for the engines to advance to 100 percent. Bill did not say anything on the radio as we watched the aircraft drop out of the sky, hit the runway, and bounce back into the air. We could hear the engines begin to spool up. The aircraft was now beginning to drift off the runway centerline over to the grass and weeds that were along the sides of the runway as Ammand struggled to keep the aircraft airborne. As the aircraft was descending lower and lower to the ground, grass, and weeds, dust began to fly up from the engines exhaust. When the aircraft was approximately a foot or so above the ground, the engines reached 100 percent, and the aircraft slowly began to climb. Ammand had managed to keep from crashing. Bill and I were both standing now and looked at each other with neither or us saying a word.

After that, as luck would have it, we noticed a blue Air Force car approaching the RSU. It was our Squadron Commander's car. This was the last person in the world that I wanted to see come out to the RSU for this event. He parked his car, got out, and entered the RSU. He said, "Dave, I heard that you were soloing Ammand today, so I thought I would come out and see how it goes." Bill and I did not say a thing as the Squadron Commander took a seat. Bill and I again could only hope for the best. This was especially true now after his first almost disastrous attempt at landing.

As Ammand came around for his second landing, I think both Bill and I truly were playing "you bet your wings." As Ammand turned final for his second landing, he did not seem to be fighting the aircraft as much. In fact, as he arrived over the threshold of the runway, he was in pretty good position. He actually touched down normally and advanced the power to take off again for his third landing. Bill and I looked at each other and did not say a word. The Squadron Commander also did not say anything.

As we waited for Ammand to come around again for his final landing, I think both Bill and I had our fingers crossed, hoping Ammand could do it again. Thankfully, his last landing was quite similar to his second landing. After Ammand's full stop landing, the Squadron Commander got up and said, "Good job, Dave," and left. Both Bill and I breathed a tremendous sigh of relief because we were going to be able to keep our wings. I think Ammand scared himself so bad on the first landing, he truly was concentrating on the last two. Those last two landings were the best landings that I ever saw him make. I also believe that God had played a part in this exercise, and this was part of God's plan for Ammand as well as me.

As a further example of how bad of a pilot Ammand was, he actually, got me into an inadvertent spin on one of our missions. He was the only student that ever did that to me as an instructor pilot. In fact, he was the only student that I ever heard of that got anyone into an inadvertent spin.

The T-37 aircraft was the only Air Force aircraft that would spin and the pilot could recover it if he recognized the spin. This was the reason we taught spins in the T-37 in order for the pilot to recognize a spin if they happen to get into one. In all other Air Force aircrafts, if you get into a spin, the normal course of action is to eject. You truly have to be a bad pilot to get into a spin.

On the particular day that Ammand got us into the inadvertent spin, we were practicing aerobatics. He was attempting to do an Immelmann maneuver. This aerobatic maneuver is when you dive the aircraft to get sufficient airspeed to pull up like you are doing a loop. Only at the top of the loop when you are inverted, you roll the aircraft to the upright position. A spin consists of stalling the aircraft and after that introducing yaw, which causes the aircraft to snap into an accelerated spin. At this point, the aircraft is no longer flying and is spinning and falling vertically like a spinning rock. The problem that Ammand had was that he misread his airspeed indicator and entered the maneuver one hundred knots low on airspeed. As he pulled up to the inverted position, the aircraft ran out of airspeed and stalled. When he tried to roll the aircraft, yaw was introduced, and the aircraft immediately snapped rolled into an accelerated spin. This now was a significant problem for us. The altitude loss during a spin is approximately ten thousand feet per minute. Because of this, when we practiced spins, we had to be above eighteen thousand feet to enter the spin. If you are not out of the spin by ten thousand feet, you are supposed to eject.

Our problem began with our entry into the spin. We were only approximately at fifteen thousand feet when we entered the spin. I had to allow Ammand time to see if he was going to recognize that we had entered a spin and to see if he could recover the aircraft. As we were passing nine thousand feet, it became totally obvious that Ammand did not have a clue as to what was happening. I calmly told Ammand to give me the aircraft so that I could save our lives. He did let go of the stick and held his hands up above his head and said, "Ammand make mistake." This was his normal response anytime that he would dazzle me with his lack of flying abilities. I took control of the aircraft. Since we were in an accelerated spin, I had to get the aircraft back into a normal spin before I could affect a recovery. I completed the spin recovery at about two thousand feet directly above the small Texas town of Spofford. God was definitely with us that day. If we had ejected, the aircraft would have crashed into Spofford with a high probability of catastrophic results on the ground. Needless to say, that was the end of that mission on that day. My only concern now was that someone from Spofford would call the base and complain about someone buzzing his or her town. No one did, so I escaped any serious problems after that.

With God's help, I did get Ammand through the T-37 program. I knew, nevertheless, that he would never be able to complete the T-38 program. The T-38 aircraft is a high-performance aircraft, and I knew it was well beyond Ammand's flying capabilities.

I was correct. After seven flights in the T-38, including his elimination flight, he was washed out of the training program. It was quite obvious to everyone that he would certainly kill himself and maybe others if he were allowed to continue.

Ammand returned to Iran and was assigned to fly the C-130 aircraft. Since he was royalty, I am sure he flew as a copilot in the C-130 aircraft for a minimum respectable period of time and after that assumed his place as third in line to the Shah.

I think my success with Ammand impressed my Squadron Commander. It was not long after Ammand washed out of the T-38 program that I was selected to become a Squadron Check Pilot. We had four check pilots for the entire squadron, and we administered check rides to all of the students as they completed each phase of training.

I was in check section for about a year when God provided me with another opportunity. The Vietnam War was building up and a number of pilots were getting assignments to go to Vietnam. When an opening occurred in the Wing Standardization/Evaluation (stan/ eval) section after one of the T-37 pilots left for Vietnam, I was selected to be his replacement. There were only three T-37 slots in the Wing Stan/Eval section. The function of the Wing Stan/Eval section was to administer check rides to all the instructor pilots on a yearly basis to ensure their proficiency. In addition, Wing Stan/ Eval pilots checked out all newly assigned instructors before they were assigned to the training squadrons.

Of the three T-37 stan/eval pilots, one was also the Wing's T-37 spin pilot. The T-37 had seven different spin modes, which also included an inverted spin. The wing's spin pilot's job was to train all T-37 instructor pilots on all seven of the spin modes. Only the normal spin mode was instructed and practiced with the students. The fact that Ammand had gotten me into that inadvertent spin probably did influence the decision for me being selected to be the Wing's spin pilot. I would instruct all newly assigned T-37 pilots on the seven different spin modes of the T-37, including the inverted spin. The inverted spin was quite uncomfortable because the aircraft would spin in the inverted position with negative g-forces (gravity forces) being felt by the pilot until recovery. Squadron instructor pilots only taught the normal spin mode but needed to be at least familiar with the other six modes that could occur.

# CHAPTER 6

# A Nation at War

THE 1960S WERE A turbulent time in America. At home, protests and social change were reshaping the country, but in military circles, the focus was squarely on Vietnam. I watched as pilots I had trained went overseas—some never to return. At Laughlin, about 50% of the instructor pilots were separating from the military as soon as they had fulfilled their commitment to the Air Force instead of waiting to go to Vietnam. Those of us that stayed in the Air Force knew that we would be going to Vietnam, since no other assignments were being received.

The Vietnam War was escalating. By 1968, the Tet Offensive had shocked the American public, and political tensions were running high back home. Yet in military circles, there was little debate: orders were orders, and duty was duty. Still, morale had an edge to it. We were training under a shadow, aware that the next set of orders could send any one of us into combat.

I didn't want to wait for the inevitable. That year, I went over to the personnel office and volunteered for Vietnam. I volunteered for fighters or forward air controller (FAC)—roles that carried some of the highest casualty rates in the war. The Air Force, nevertheless, had other plans. My name sat on the volunteer list for months.

The Air Force told me I was needed more as an instructor at that time, and I remained frozen stateside for another two years. But I knew my turn was coming. Every pilot in those days understood that Vietnam was inevitable. We trained with that reality in mind, honing our skills not to stay alive but to ensure the safety of those we would one day be tasked with protecting. During that time, nevertheless, the first 19 pilots from Laughlin that left did not make it back. They were killed in action or were shot down and taken prisoner. If I had not been frozen when I volunteered, I would have been part of those 19 pilots. This is why I know that God was guiding my life, because he had other plans for me.

# CHAPTER 7

# Preparing for Vietnam

In December 1970, the long-awaited orders came: I was assigned as an O-2A Forward Air Controller with fighter qualification. While part of me had hoped for a fighter assignment, I understood that FAC duty was essential. FACs were the eyes of the battlefield, coordinating air support and saving countless lives. It was dangerous, but it was necessary. And I was ready.

The next six months were a whirlwind of training. I went through fighter qualification at Cannon Air Force Base in New Mexico, followed by water survival training at Homestead AFB in Florida and after that basic survival training at Fairchild AFB in Washington State. Finally, I trained in the O-2A aircraft at Hurlburt Field in Florida. By the time I boarded a plane for Southeast Asia in June 1971, which included an intermittent stop in the Philippines for jungle survival training, I arrived in Vietnam, confident in my training but fully aware of the risks.

Vietnam would test every skill I had learned, every ounce of courage I possessed, and every word of Scripture I had ever heard as a boy at my grandmother's kitchen table. I didn't know it yet, but those lessons of faith and resilience would one day save my life.

# CHAPTER 8

# The Making of a Military Aviator

By the late 1960s, I had already logged thousands of hours in the cockpit as a T-37 instructor. Flying was no longer a skill; it was muscle memory, instinct, and part of my identity. Yet there was a part of me that felt untested. Training pilots was rewarding, but as the war in Vietnam escalated, I knew that wearing the uniform meant stepping into harm's way.

At Laughlin Air Force Base in Del Rio, Texas, we lived by schedules and precision. Days started before dawn with preflight briefings, followed by sorties in the hot Texas sun. The T-37 Tweet was no powerhouse—more like a stubby-winged schoolhouse jet—but she was forgiving enough for young pilots. My job as an instructor pilot, check pilot, stan/eval pilot and T-37 spin pilot provided me with a tremendous amount of experience that would benefit me greatly in the future. This was all part of God's plan for me. I would have never gotten the type and level of experience if I had received a T-38 instructor slot after graduation.

# CHAPTER 9

# Fighter School and Survival Training

Leaving Laughlin AFB was bittersweet. My years as an instructor had made me part of a tight-knit community of pilots and families, but my heart was ready for the challenge ahead. The next six months became a blur of intense training.

At Cannon Air Force Base in Clovis, New Mexico, I completed Fighter Qualification School. Cannon Air Force Base was the home of the 27th Tactical Fighter Wing. The base was the home of four F-111D Tactical Fighter Squadrons and one Fighter Qualification Squadron. This squadron was the 4429th Combat Crew Training Squadron. For FAC Fighter Qualification training, to save money, the Air Force modified several T-33s that were being replaced by the T-38 for UPT training. The modifications included, installing two 50 caliber machine guns in the nose of the aircraft and adding bomb dropping capabilities. These modified T-33s became AT and BT-33s. These aircraft were quite similar to the F-80 fighters used in Korea.

It was at Cannon AFB, where each Friday during a safety briefing, a list of fighter losses in Vietnam for the previous week were shown. The listed always had anywhere from 4 to 6 pilot names on it. It seemed that I recognized at least one name and sometimes more of the names at ever meeting. Like I said the first 19 pilots that left Laughlin did not make it back. In fact, the first pilots to make it back to Laughlin occurred in December 1970, when I received my orders. We got back two F-4 pilots and one F-105 pilot.

Though my primary assignment would be flying the O-2A Skymaster—a slow, twin-engine propeller plane that was a civilian aircraft that had been converted into a war machine—I needed to be fighter-qualified. The Army required FACs supporting ground troops to have fighter experience so they could better coordinate strikes from fast-moving jets. That meant I had to train like a fighter pilot, even if I'd be flying a bird with a top speed barely over 120 knots.

Cannon AFB taught me humility. The O-2A was nothing like the supersonic T-38 I'd flown as a student or even the AT/BT-33. It had no ejection seat, no heavy armor, and no defensive systems. It only carried a couple of smoke rockets pods with 7 rockets each mounted under its wings to mark targets. The cockpit smelled faintly of oil and sweat, and its thin aluminum skin wouldn't stop a bullet. I realized quickly that my survival wouldn't come from the aircraft's capabilities but from my own skill, training, and faith.

After Cannon came water survival school at Homestead Air Force Base in Florida. We spent days in the pool or the ocean learning how to escape from being under a parachute in the water. If you eject and land in the water, the parachute will usually clasped on you as it deflates. This will result in you being under the chute. You have to know how to get out from under the chute or you could drown. We also learned how to stay alive on minimal rations. Instructors simulated real-life emergencies, flipping us upside down in rafts and dragging us through simulated rough surf while we tried to deploy rafts.

Next came basic survival school at Fairchild Air Force Base in Washington State. There, we learned the grim realities of surviving in hostile territory. Instructors taught us to blend into our surroundings, avoid detection, and find food and water in dense vegetation. We practiced evading simulated enemy patrols and spent nights shivering (it was February) in makeshift shelters. As part of this training, there was also a portion simulating a Prisoner of War (POW) exercise. This was quite realistic. Survival training wasn't about skills; it was about mental toughness. The message was clear: in Vietnam, rescue was never guaranteed.

At Hurlburt Field in Florida, I learned to fly the O-2A. The Skymaster was a humble aircraft compared to the sleek fighters that dominated Air Force lore, but in Vietnam, FACs were the lifeline between ground troops and air support. We flew low and slow, identifying targets, marking them with smoke rockets, and coordinating strikes. The mission demanded precision flying, split-second communication, and nerves of steel. And finally, during the intermittent stop in the Philippines for jungle survival training, we were taught how to survive and escape and evade in a jungle environment. By June 1971, I was ready—as ready as anyone could be.

# CHAPTER 10

# First Steps into a War Zone

MY FLIGHT TO VIETNAM was uneventful, but the welcome was anything but. As the plane arrived at Cam Ranh Bay, I became a little confused. I had finally arrived in the place I'd trained for months to face. Our aircraft landed and rolled to a stop, not at the terminal, but in a "hot brake" area—a remote part of the tarmac. Confused, I followed the other passengers as we were herded off the plane. When the flight crew deplaned, they did not go to the busses that had been sent to pick us up. They went to the tail of the aircraft. I decided to follow them. That's when I saw it: a clean, ragged hole punched straight through the horizontal stabilizer. We had taken a 50-caliber round on approach, and none of us had even known it. That was my first lesson about Vietnam: danger didn't announce itself. I did think to myself that this was going to be a very long year.

Cam Ranh Bay was the home of the 504th Tactical Air Support Group. It was called a Group and not a Wing because with Wings, all assigned squadrons are assigned to that base. For a Group, some of the assigned squadrons are assigned to other bases. The squadrons assigned to the 504th was the 19th Tactical Air Support Squadron (TASS) and was at Cam Ranh Bay. The 20th TASS was assigned to DaNang Air Base and the 21st TASS was assigned to Tan Son Nhut Air Base. The 23rd TASS was assigned to a base in Thailand.

Vietnam had been divided up into four Military Regions (MR), MRI, MRII, MRIII, and MRIV. The 20th TASS was responsible for MRI and the 19th TASS was responsible for MRII. The 21st TASS was responsible for MRIII and half of Cambodia. The Navy was responsible for MRIV.

# CHAPTER 11

# Working My Way North and My New Home

AFTER A NIGHT AT Cam Ranh Bay, I was assigned to the 20th Tactical Air Support Squadron, based in Da Nang. I spend a couple of days there and after that was sent north to Hue Phu Bai. Phu Bai was the main operating base for the 2nd Brigade of the 101st Airborne Division. Throughout their operating Area of Responsibility (AR), the 2nd Brigade also established forward operating locations called "Fire Bases" (FBs). Units from the 2nd Brigade would be deployed to these FBs for varying lengths of time. From these FBs, combat operations would be conducted around the area of each FB. These FBs each had names and usually were located on top of mountains. The Army would send out reconnaissance patrols from these FBs to identify enemy targets. They would provide this intelligence of possible targets back to Phu Bai for consideration of engagement. The Tactical Operations Center (TOCs) for both the Air Force and the Army were co-located to allow for smoother operations in support of combat activities. This combined center for both the Army and Air Force's TOCs was called the Tactical Air Support Center (TASC). The Air Force FAC operations along with the Army helicopter operations were controlled out of this center. If the Army felt that they could handle the target with their helicopter assets, they would service the target request. If the target was more of a problem, they would request the Air Force to service the target. If a "troops-in-contact" (TIC) situation occurred at any of these FBs, we all would do whatever was necessary to address the situation.

For those targets that were not mobile in nature, we would order air strikes for the next day. We would order these air strikes through the Direct Air Support Center (DASC), which was located at Da Nang AB. The DASC at Da Nang had tactical air support responsibility for all of MRI or I Corps. If tactical air support was needed immediately, we would either launch a FAC or have one of our airborne FACs address the situation with air support. We always had at least two FACs airborne at all times to assist any Army patrol that might run into trouble. These FACs were also conducting reconnaissance in the AR looking for targets as well as being available if called upon.

Additionally, some of the Ho Chi Minh Trail infiltration routes from North Vietnam through Laos had entry points into South Vietnam in our AR. We would interdict these entry points each day before sunset. Our goal was to close down the trails into South Vietnam or at least slow down the North from bringing supplies into the South each night to resupply the Vietcong (VC or "Charlie"). This was accomplished either by air strikes or the use of the Army's 175 artillery tubes. After that, each morning we would do a "first look" at each of these interdiction points to see how much truck traffic had occurred during the night.

Our daily FAC operations out of Phu Bai consisted of supporting the Army with close air support missions, directing scheduled air strikes on fixed targets, interdiction missions and reconnaissance missions.

# CHAPTER 12

# First Combat Missions

AFTER MY ARRIVAL AT Phu Bai, the Air Logistics Officer (ALO), a Major Robbins (Robby), gave me two day AR orientation sorties. I was now cleared to fly day missions in support of the Army. A quite interesting thing happened on the 2nd or 3rd mission that I flew. I was out in the western part of the AR, in the A Shau Valley, when a voice came up on my ultrahigh frequency (UHF) radio. The voice spoke in broken English and welcomed me to Vietnam. My call sign at that time was "Bilk 28." This voice said, "Welcome to Vietnam Bilk 28, we are going to shoot you down." This was a little unsettling to say the least. Once again, I thought to myself, this is going to be a very long year.

When I got back to Phu Bai, I asked Robby if this was a normal occurrence. He responded by saying, "Oh yes, Charlie welcomes all of us FACs to Vietnam." It also told me that our enemy had a quite significant intelligence network in place. They knew I was new to Vietnam, and they knew my call sign. They also knew our radio frequencies. One must wonder how they got this type of information. One also had to be impressed that they got the information so fast.

I now understood why Robby had stressed that I should always use a different departure route after takeoff when departing Phu Bai. It was obvious that "Charlie" was monitoring our takeoffs and landings as well as our departure routes. Being random in departure was the best way to avoid being predictable, which could lead to being a good target on takeoff. I never forgot this lesson, especially after my arrival on the Continental flight and the 50-caliber round through the horizontal stabilizer on landing. This lesson served me well in Vietnam.

The O-2A has two engines in a push/pull configuration. The pull engine is on the front of the aircraft while the push or rear engine is located behind the cockpit between the two booms of the tail section. With both engines operating, one can achieve altitudes up to approximately 8,000 to 10,000 feet for sustained level flight. If you lost the rear engine, you could only maintain a maximum altitude of approximately 1000 feet for sustained level flight. During my tour in Vietnam, I lost the rear engine on three different missions. I only lost engines twice during the rest of my 25-year Air Force career.

The first time I lost the rear engine in Vietnam occurred on a mission after "Charlie" had welcomed me to Vietnam. I was out in the A Shau Valley on a normal reconnaissance mission when I lost oil pressure on the rear engine. The engines on the O-2A automatically shut down if oil pressure is lost. The A Shau Valley is surrounded by mountains that range from 3000 feet to 5000 feet in altitude. There is only one pass out of the A Shau Valley where the ground level is below 1000 feet.

When I lost the engine, I was near the southern end of the Valley; and the pass is near the northern end. As I began the mandatory descent to a level that I could maintain level flight, I headed back to the only area that would allow me to exit the Valley. As I proceeded through the pass to the east out of the valley, I noticed 50-caliber traces outside the cockpit coming from above me from the top of a mountain, and I wondered if "Charlie" was going to make good on his welcoming greetings of shooting me down. Tracer rounds are used every fifth round. I saw a number of tracers coming from above me and passing below. Thankfully, I made it through the pass and back to Phu Bai without being hit. That was my first mission on which I took ground fire but definitely not the last. Again, I thought to myself this truly is going to be a very long year.

It was on August 8, 1971, 16 days after my first flight in Vietnam, when I was awakened at 2300 hours (11:00 P.M.) by Maj. Robbins

(Robby). Robby told me that we had a very critical "troops in contact (TIC)" situation at Fire Base (FB) Sarge which was located in the most western section of Quang Tri Province, in the 1st Brigade's AR. I reminded Robby that I had not had my night checkout mission and that I had not flown in Quang Tri Province area since that AR belonged to the FACs at Camp Evans. He responded by saying "Yes, I know, but the ALO at Camp Evans called and said none of his pilots were in crew rest to fly." Robby had responded to him that none of his pilots were in crew rest either. The problem for not being in crew rest was due to the fact that after flying operations had ended for the day, all of the pilots at both locations had been relaxing and drinking beer. I do not use alcohol. The ALO at Camp Evans was my old Flight Commander at Laughlin AFB in Del Rio and knew that I was supposed to arrive soon in Vietnam and that I did not drink alcohol. He asked Robby if Scheiding had arrived yet? Robby said "yes, but I had not been checked out yet at night." My old Flight Commander said no problem — send Scheiding, he can do it. After that, Robby said "The situation is critical, and you have to go." I guess he figured that since I was a mid-range Captain and I previously been an Instructor Pilot (IP), I could handle it. I was also the next ranking officer at this location. I immediately got up and rapidly got dressed for flight. Once again, the thought crossed my mind that this is going to be a very long year.

While I was getting dressed, Robby got our maintenance personnel to get an aircraft ready for flight. Robby met me at the aircraft and said that I would not have any problems finding FB Sarge. He said, " fly up the coast to Quang Tri City, look west, and where you see all of the action going on — that's FB Sarge." Robby was correct. I had no problem finding FB Sarge. It looked like a 4th of July celebration with fireworks. I arrived on station 45 minutes after being awakened by Robby.

When I arrived on station, the Army was providing tactical fire support with their 175 artillery batteries. I contacted the ground commander at FB Sarge, and he provided me with all of the information I needed to take over control of the close air support. I then contacted the 175-artillery battery commander and began directing his fire support on the areas where the North Vietnamese

Army (NVA) and/or the Vietcong (VC) were attacking the FB. FB Sarge was being attacked on three sides. I immediately knew that I would need more fire power, so I contacted our Direct Air Support Center (DASC) at Da Nang and requested any air support that was available for a "troops in contact" (TIC) situation. The Air Force normally had AC-130 gunship air strikes in progress each night over the Ho Chi Minh Trails in Laos, which were not far from our location. I also knew that we had F-4 fighters on alert at Da Nang.

It was not long before an AC-130 gunship checked in with me from off the trails. I held up the 175s and began directing the AC-130 on the areas of most concern. When the gunship was "Winchester" (out of ammo) he departed for home. After that, I returned to the 175s to keep ordinance falling on the "bad guys" while I waited for additional air support.

It was not long before the first set of F-4s from the Gunfighter Squadron at Da Nang checked in with me. I put these F-4s in close to the FB as the "bad guys" were getting quite close to the FB's perimeter. During the next three and a half hours, I was provided with three additional sets of fighters and another gunship. So, for four and a half hours, I directed four sets of fighters, two gunships, and 175 artillery tubes in support of FB Sarge, which did result in finally breaking the "troops in contact" situation. After four and a half hours "Charlie" decided he had had enough and retreated back into Laos. That was quite good for FB Sarge and good for me, since I was getting low on fuel and I was almost out of marking rockets. We had turned back "Charlie" and kept him from overrunning FB Sarge that night. I returned to Phu Bai as the sun was beginning to rise.

Robby met me at the aircraft with a big smile on his face. He had been monitoring the action from our Tactical Operations Center (TOC) and was fully aware of what had happened. As I got out of the aircraft, I asked Robby if I should call Da Nang and tell them that I had my night checkout. He smiled and said, "By all means."

The next day, FB Sarge called our TOC and provided a bomb damage assessment (BDA) for the previous night's operation. They reported 258 "bad guys" killed in action (KIA) and also said that "Charlie" had dragged off many more.

I was awarded the Distinguished Flying Cross (DFC) medal for that mission. I was happy that I was able to support our troops on the ground that night. I also learned some quite valuable lessons. I learned how vicious war actually is and what intense automatic weapons ground fire looks like. Being a fighter pilot and going fast, dropping bombs and firing guns is fun, but there is no greater feeling than knowing that you were responsible for providing the means that actually saved the lives of our troops on the ground. I knew my actions had saved many lives that night at FB Sarge by me doing my job.

# CHAPTER 13

# Life in The Shadow of Danger

THE FIRST THING I noticed at Hue Phu Bai was how different life was from my previous bases. In the U.S., pilots are required to have air-conditioned quarters, clean dining facilities, and reliable infrastructure. At Phu Bia we lived like soldiers in "hooches." The "hooches" were wooden huts with thin walls, housing four pilots apiece. Some had air conditioning; others didn't. Electricity was rationed, available only for a few hours each evening. The bathrooms were little more than outhouses with 55-gallon barrels beneath the seats, emptied daily by Vietnamese workers. Showers offered only cold water, and we walked to them along plywood pallets sidewalks to avoid the mud.

But despite the rough conditions, FAC morale was surprisingly high. FACs were a tight-knit community, and the camaraderie we built there was unlike anything I'd experienced before. Life at Hue Phu Bai was a strange mix of boredom and terror. On some days, the biggest challenge was dealing with the relentless heat or scavenging for cold soda at the Post Exchange. On others, we scrambled under mortar or rocket fire, diving into sandbag bunkers as rockets screamed overhead.

Despite the constant danger, we adapted. Humor became a coping mechanism. Pilots decorated their hooches with makeshift furniture, bartered for creature comforts, and built a sense of home in a place where home was a fragile illusion.

But even in those early days, the war was personal. Names of fallen pilots circulated quickly, and every mission carried the weight of uncertainty. We flew knowing that at any moment, an ambush could bring us down. For the 10 years of the Vietnam War, a total of 238 FACs were killed in action. That is essentially 2 pilots a month for 10 years.

Still, there was purpose in what we did. FACs saved lives every day. We were the link between soldiers on the ground and the air power that could mean the difference between survival and annihilation. That responsibility gave meaning to every mission. The importance of the FAC mission was recognized. Of the 12 Medals of Honor awarded during the Vietnam War, 2 went to FACS.

# CHAPTER 14

# Faith in the Cockpit

EVEN IN MY FIRST few weeks, I felt the presence of God in ways I had never experienced before. My faith, instilled by my parents and grandmother, became more than a belief—it became a lifeline. I couldn't have known it after that, but this deepening faith would one day save my life. In the months ahead, I would hear a voice—clear, unmistakable, and divine—telling me not to fly. That moment would mark me forever.

For now, I focused on learning my craft, building relationships with the Army units I supported, and surviving each mission. The war was beginning for me, and it would test every part of who I was: a pilot, a leader, a husband, a father, and a man of faith.

# CHAPTER 15

# Normal Operations

THE DAILY MISSIONS BECAME routine. I flew combat patrol, interdiction and reconnaissance missions. In addition, the Army would conduct search and destroy missions. We would also participate in these types of missions which consisted of the Army identifying a relatively large area as a "free fire zone". These areas were identified as suspected enemy held areas, and all of the required clearances would be obtained for a specific period of time prior to takeoff. These missions consisted of what the Army called "Pink Teams". These teams consisted of a "Little Bird," two Cobra gunships and a "Command and Control" (CC) Huey helicopter. A FAC would also accompany these teams on the search and destroy missions into the designated "free fire zone."

These missions consisted of "Little Bird" flying on the deck above the treetops, trolling for enemy gunfire. The crew of this helicopter consisted of a pilot and a gunner sitting at the door with his feet on one of the landing rails. In one hand, the gunner had an M60 machine gun; and in the other hand, he held a smoke grenade. Flying above "Little Bird" were two Cobra gunships that were flying in circles on either side of "Little Bird's" flight path. These circles were coordinated to ensure that at least one of the Cobra gunships was in a position to respond if "Little Bird" took fire. Above these three helicopters, the Huey Command and Control (CC) helicopter was positioned. His function was to be in charge of the mission. Above these four helicopters, in this moving stack of aircraft, was the FAC. When "Little Bird" took fire, the gunner would immediately throw the smoke grenade and start firing the M60 machine gun. When the smoke grenade ignited, the gunships would immediately roll in on the target as "Little Bird" exited the area. If the target turned out to be too much for the gunships to handle, the target was turned over to the FAC for engagement. The FAC would order an air strike through the TASC who in turn would contact the DASC at Da Nang for support. The target would then be serviced with fighter aircraft. This was a quite effective way to neutralize targets and prosecute the war.

I did have an interesting mission on one of the daily "first look" sorties to check out the interdiction points. I discovered a truck that had slid off of the trail and was stuck. What probably happened was that after the Vietcong (VC or "Charlie) had cleared the trail, this truck was probably trying to beat sunrise. It most likely took "Charlie" longer to open the trail after our closure the night before, and this truck was going too fast as it was trying to get through this interdiction point before sunrise. Once past these points, the trucks would disappear into the triple canopy jungle.

On this day, one of the truck's back wheels slipped off of the trail; and the truck became stuck in the open area of the interdiction point. I spotted the truck, and the driver was trying to conceal the vehicle by shoveling dirt on top of it. He was never going to be able to cover up the truck, but at least he was trying.

I contacted our TASC who contacted the DASC at Da Nang and requested air support. About 20 minutes later two F-4s from the Gunfighter Squadron at Da Nang checked in with me. I briefed them on the target and told them that I would mark it. Marking the target probably was not necessary since the truck was quite obvious against the bare soil from the repeated closures each night. It was, nevertheless, always fun to shoot rockets, and the smoke would assist the fighters by giving them the existing wind conditions.

When the smoke rocket detonated, the truck driver disappeared into the jungle. I decided to have the fighters drop one bomb at a time since I had not observed any ground fire coming from the target area. Lead dropped his first bomb close to the truck but did not destroy it. Two rolled in; and when he released his bomb, it landed on the covered bed of the truck and exploded. The truck tumbled down the side of the mountain and was no longer a concern. The fighters wanted to know where I wanted the remainder of their bomb load placed.

Since we felt that the North Vietnamese Army (NVA) or the Vietcong (VC or "Charlie) used heavy equipment, like bulldozers or front end loaders to clear the trails each night, I decided to try and get lucky. We felt that this heavy equipment was probably located near each of the interdiction points and concealed by the triple canopy jungle. I directed the fighters to drop their remaining bombs in the jungle area near the trail and the interdiction point. Who knows, we may have gotten lucky and actually found the needle in the haystack.

While at Phu Bai, I also learned another valuable lesson. I learned to never trust any of the Vietnamese working on base. Supposedly, these workers had all been vetted, but how can one truly be sure where these workers' allegiance actually lies.

This lesson was driven home one night around 0200 hours (2:00 A.M.) when the base came under a rocket attack. In addition, "Charlie" was probing our perimeter razor wire fence. After a short time of small arms fire near the wire, the probe ceased, as did the rockets. The next morning two of the individuals who had been doing the probing had been killed and were lying in the wire. One of these individuals happened to be one of our Vietnamese barbers on base. He cut our hair during the day; and after that at night he became "Charlie" trying to kill us. You never truly knew who the enemy was.

I also saw the good and the bad of the Army while at Phu Bai. The good was displayed by the actions of many of the helicopter pilots. It was on one of the search-and-destroy missions that I witnessed the loyalty and the strong bond that is created between fellow soldiers during war.

On this particular mission, "Little Bird" was shot down and crashed on top of the triple canopy jungle. When it crashed, the small chopper seemed to roll like an egg on top of the triple canopy. Since "Little Birds" are shaped like eggs, this crash pattern was not unexpected. What was unexpected was when the Huey CC bird descended to the crash site to check on the crew, at about 15-20 feet above the crash site while in a hover, one of the crew members of the CC ship jumped out of the helicopter without a parachute. He landed on top of the triple canopy jungle and made his way over to "Little Bird" where he pulled both crew members out. I watched in amazement at this unselfish and brave act. Unfortunately, both crew members had been killed in the crash. When asked later why he had jumped out of the CC Bird, this individual merely responded saying, "I had to. The pilot of 'Little Bird' was my buddy." That is what dedication, bravery, and loyalty is all about. The bond developed between military members in wartime is extremely strong. One has to respect it.

Another good example of the dedication, bravery and loyalty of these young 18-and 19-year-old helicopter pilots was displayed when an Army patrol team out in the A Shau Valley west of Phu Bai became pinned down by enemy fire. The Army TOC launched a number of Cobra gunships to provide close air support to these troops.

The report of "troops-in-contact" (TIC) came in after sunrise. A number of gunships were launched to address the situation. At approximately 0700 hours (7:00 A.M.), one of the gunships was shot down. A rescue Huey Medevac helicopter ("Dust Off") was able to retrieve the pilot and gunner and bring them back to Phu Bai. The pilot immediately went to the Flight Surgeon and requested to be cleared for flight. Fortunately, he was not injured, and the Flight Surgeon did clear him for flight. He immediately came back to the TOC and requested to go back out to the site since the situation had not been resolved. The Army Brigade Commander authorized his return to the battle in another gunship.

At approximately 1200 hours (12:00 P.M.), he was shot down again; and again "Dust Off" was able to rescue him unhurt. He again went to the Flight Surgeon and requested to be cleared for flight. The Flight Surgeon once again cleared him, and he returned to the TOC. This time the Brigade Commander sent him out in a Huey Command and Control (CC) Bird to reenter the battle area.

As bad luck would have it, at approximately 1600 hours (4:00 P.M.) his CC Bird was shot down. He and the crew were again rescued unharmed and returned to Phu Bai. This time after the Flight Surgeon cleared him for flight, the Brigade Commander refused to let him go back out to the site. The Brigade Commander said, "Three choppers in one day are enough. I can't afford you flying anymore today." The young Warrant Officer did not like this answer, but I think he understood. I do not know if this was a record or not, being shot down three times in one day, but it did exhibit the type of courage and bravery that many of these young Army helicopter pilots had.

In addition, many of these young Warrant Officers had a goal of getting 1000 hours of flight time during their year in Vietnam. If they did, they could return to the States and get a good paying job as a helicopter pilot for a radio station news department or a job flying helicopters transporting patients. They would do anything to get flight time.

I observed an example of one of these young pilots trying to get as much flying time as possible. I happened to be getting my annual flight physical when a young Army Warrant Officer pilot poked his head into the Army's Flight Surgeon's office and asked if he could be released from DNIF status (Duty Not Involving Flight). This is a status all of us pilots hate. The Flight Surgeon looked at him and asked, "How high can you raise your arm?" His right arm was in a sling. The younger officer slipped off the sling and tried to raise his arm. He could only raise it up to about shoulder level. The Flight Surgeon looked at him and said, "Not today son." The young pilot put the sling back on and went on his way. As he departed, he said he would be back tomorrow.

The Flight Surgeon then turned to me and said that he had taken 12 or 13 pieces of helicopter canopy plexiglas out of the young pilot's arm the day before when he had been shot down. These young chopper pilots had no fear, and they felt that they were bulletproof. I guess that is why the Army liked these young pilots. These young pilots had quite high morale while the Army ground troops that manned the FBs for long periods of time had quite low morale.

In addition to the good I also witnessed the bad. During this time period, outside of the helicopter pilots, the Army in general was experiencing quite low morale as well as a significant drug problem among its enlisted members.

As I previously mentioned, our Tactical Operations Center (TOC) was co-located with the Army's TOC for its helicopter assets. The Army manned their half of the TOC 24 hours a day. We manned our half only during the time we had flight operations going on. Each night, nevertheless, I had to go over to the TOC to get the next day's Fragmentation Order (FRAG) which specified the next day's scheduled air strikes in our AR. Since I was the second ranking officer, Major Robbins (Robby) had put me in charge of scheduling. It was my job to schedule our daily flight operations. The TOC was located in a secured bunker, which used a daily password for entry for security purposes. In order to enter the TOC, the daily password had to be used. Initially, when I took over flight scheduling and I would go to the TOC at 2200 hours (10:00 P.M.) each night to get the FRAG, I did not notice any potential problems. As time passed, I started noticing small empty plastic vials lying next to the PSP walkway leading to the TOC. At first it was one or two of these empty vials. Not knowing what these vials were, I asked. I was told that these vials, which were in two sizes, were nickel and dime bags for drugs. After that, this became a big concern for me. The Army leaders seemed to be aware of the problem but seemed to turn their heads as if not wanting to know. I, of course, had a great concern since I had to enter the TOC each night at 2200 hours (10:00 P.M.). To make matters worse, the number of empty vials was increasing over time.

To address my concern and even though the Army personnel manning the TOC knew that I would be coming, I would make as much noise as I could while approaching the TOC. I would whistle or walk heavily on the PSP to let them know that I was approaching. This gave them enough time to put away anything that they did not want me to see and anything that I certainly did not want to see. I also hoped that they would remember the password for the day. The number of empty vials never decreased during the time I was at Phu Bai.

I was also there when an incident of fragging occurred. Fragging is when someone would open one of the doors on either end of a living quarter's hooch and toss a live hand grenade inside. It was usually an action being accomplished by one Army individual against another. It usually involved an enlisted member fragging an officer for whatever reason. I am sure that drugs were a contributing factor to this type of activity. Fortunately, the individual that was fragged, while I was there, was not killed; however, no one was ever identified that I know of as the Frager. Other than for the helicopter pilots, morale was not high in the Army during the latter part of 1971.

Another one of a kind mission that I flew in support of the Army involved the creation of an instantaneous five-ship landing zone (LZ). The Army had decided to enter an area that was known as a VC stronghold area and the Army had not paid too much attention to it in the past.

The Air Force had a few 15,000-lb bombs available, that when dropped out of the back of a C-130 aircraft, would create an instantaneous five-ship helicopter LZ. A six-foot fuse extender was attached to the nose of the bomb, and after that it would be dropped out of a low-flying C-130 aircraft with the use of parachutes. Three large parachutes would slow the descent of the bomb and orientate the bomb nose down as the C-130 exited the area. I had never seen one of those bombs dropped before.

I was briefed before takeoff as to where the Army wanted this LZ to be created. After airborne, I made contact with the C-130 aircraft and directed him into the target area. I marked the location with a smoke rocket and confirmed with the C-130 that they had the location. After that, I climbed up to an altitude of approximately 4500 feet and moved away about four to five miles from the target area. Based on my experience with 2000-lb bombs, I felt that this would be sufficient. I could not have been more wrong.

I watched as the C-130 started its run toward the target area. He was flying at an altitude of about 2000 feet and was well below me. I watched as the rear cargo door opened and the small drag chutes deployed as the C-130 approached the target area. These drag chutes pulled the 15,000-lb bomb out of the aircraft, and after that, the three large main chutes deployed to slow the rate of descent of the bomb. The bomb descended nose down while being held up by the three large chutes. This nose-down descent allows the six-foot fuse extender and the fuse to contact the ground first when the bomb is still six feet above the ground. The blast from this bomb creates an instantaneous five-ship LZ without creating much of a hole.

As I watched the bomb slowly descend and the C–130 move out of the way of the bomb blast, I felt comfortable with my position. As the bomb detonated six feet above the ground, I could see the shock wave and the blast effect, which looked like the pictures of nuclear weapons going off when these weapons were being tested. When the shock wave from the blast reached my altitude and distance away, my aircraft and I were turned every way but loose. I recovered the aircraft inverted in a 25 degree nose low attitude. I was quite lucky that no structural damage had occurred to the aircraft. I learned a very important lesson about 15,000-lb bombs—they are big! The Army was quite happy with their new LZ.

# CHAPTER 16

# Change in Responsibilities

I HAD BEEN AT Phu Bai for about two months when the 504th TASG received a message from the O-2A Training Wing at Hurlburt Field in Florida where all of the O-2A pre-Vietnam flight training was accomplished. This message was an inquiry about what the 504th TASG thought about the quality and effectiveness of the training being provided to FACs on their way to Vietnam. The 504th TASG asked all three of its squadrons to provide input to them to assist them in their response.

Each TASS was to ask a newly arrived FAC for his opinion on the quality of training he had received and how well it prepared him to accomplish the mission in Vietnam. My ALO asked me to respond since he knew that I had previously been an Instructor Pilot (IP) for Undergraduate Pilot Training (UPT) for the Air Force. He felt that since I had recently completed O-2A training, and with my IP experience, I would be the best person to evaluate the quality of training being provided in Florida and its effectiveness in Vietnam. This was especially true since most of our FACs were young Lieutenants and did not have much experience to begin with.

It probably was unfortunate for the Hurlburt Field Training Wing that I was selected to provide input. I had not been impressed with the O-2A training that I had received there. I felt that a number of things were lacking in my preparation for combat operations in Vietnam. Having been an IP for four years before Vietnam, I had the experience to overcome these shortcomings. I did, nevertheless, have a major concern for the recently UPT graduates that received FAC assignments upon graduation. I did not feel that these young pilots had the experience to overcome some of the shortcomings that I had experienced during O-2A training.

One of my major concerns about the O-2A training that I received was due to a condition that was actually beyond the control of the Training Wing at Hurlburt Field. This had to do with the physical location of the O-2A training being conducted in the Florida panhandle. This physical location made it almost impossible to teach topographic map reading skills for use in Vietnam. Topographic map reading was probably the most essential training requirement needed to be an effective FAC in MRI and MRII in Vietnam.

Hurlburt Field is located on the Florida panhandle next to the Gulf of Mexico. There essentially is no topographic relief present there for hundreds of miles. In addition, this area is heavily populated with roads, highways, towns and many more man-made features along this coastal area. This training area was totally different from the topography in Vietnam, especially in MRI and MRII. These two regions represented two thirds of the country where Air Force FACs were operating.

In these two regions, the terrain is primarily mountainous and/or covered with triple canopy jungle. There are very few roads or man-made features in the mostly uninhabited area where "Charlie" was located. In MRII, this region was called the Central Highlands, and it contained the highest mountain peak in Vietnam at a height of 8,524 feet. It was impossible to teach map reading skills to young pilots in Florida for use in Vietnam in MRI and MRII.

I also knew that map reading was not emphasized in UPT since this skill is not truly required for most of the missions in the Air Force. Topographic map reading skills, nevertheless, was mandatory for FACs in MRI and MRII. The FACs in MRIII and Cambodia did not have as much of a disadvantage since this region was mostly flat and more densely populated. This area had many roads and other man-made features for use. It was somewhat comparable to the terrain around Hurlburt Field. This concern was more of a problem for the Air Force rather than for the Training Wing at

Hurlburt Field. The Wing did not have any choice as to where the O-2A training was to be conducted.

There was a good aspect for conducting the O-2A training at Hurlburt Field. This was due to the fact that the Air Force had a number of fighter aircraft bases and ranges nearby, which allowed for excellent training of controlling fighter aircraft as a FAC. There were fighters based at McDill AFB in Tampa as well as Eglin AFB at Fort Walton Beach in Florida.

My second major concern was more of a personal problem for me. The IP that I was assigned to for O-2A training was an old Major who had been passed over for Lieutenant Colonel (LTC) a number of times and was putting in his time prior to retirement. Many times, my training missions with this IP consisted of takeoff and flying directly to the coast to see where the fish were running so that he could go fishing after he finished for the day. We would make two passes up and down the coast between Fort Walton Beach and Pensacola looking for fish. He would then have me take him back to Hurlburt Field where he would get out, and I would go solo to complete the mission.

Having been an IP myself, this did not impress me. I was doing a lot of self-instructing during my O-2A training in Florida. This was O.K. for me because of my experience; nevertheless, it would not be satisfactory for a new pilot right out of UPT. I knew that I was not his first student, and I knew I would not be his last. This type of training, in my opinion, would be totally inadequate to prepare a young pilot to be a FAC in Vietnam.

Robby forwarded my response to the 20th TASS who forwarded it on to the 504th Group for their use in preparing their response to Hurlburt Field. Two things happened after I provided my input to the 504th Group. The first thing was that I was asked by the 504th Group if I would be willing to go to Nha Trang AB and set up an in-country training program for all new FACs arriving in Vietnam. They felt that with my background as an UPT Instructor, I was the perfect one to set up such a program. They seemed to agree with me that the O-2A training at Hurlburt Field was inadequate, and they decided to do something about it. I agreed, and after two months at Phu Bai with the 101$^{st}$ Airborne, I was on my way to Nha Trang to set up a new in-country FAC training program.

The second thing that happened was that after I got to Nha Trang, a full Colonel from Hurlburt Field arrived to discuss my concerns about their training program. So there I was, a mid-range Captain telling this full Colonel how he could improve his training program. Usually, it is the full Colonel telling the Captain how the cabbage is eaten.

Air Force must have felt that my concerns were serious enough to send a full Colonel, in person, to Vietnam to listen to my concerns. I was totally surprised, but impressed, that those in a position to do something about my concerns that could mean life or death for a young pilot, were willing to listen and make changes. By the time I left Vietnam, I did see a lot of improvement in the quality of the new arriving FACs. The stateside training program had been improved.

# CHAPTER 17

# Conducting In-Country Training at Nha Trang

It was October 1971, and I was now on my way to Nha Trang Air Base. Nha Trang is located approximately 300 miles south of Phu Bai in MRII on the coast. It is situated in a valley surrounded by mountains. The 21st TASS had an FOL at this location, which provided tactical air support for MRII. I was now assigned to the 504th TASG since I was going to be providing the training for all new arriving FACs. These new FACs would be assigned to one of the three TASSs after completing this new in-country training program.

I reported to a Lieutenant Colonel (LTC) MacPhearson at the 504th Group. He also had been an UPT Instructor Pilot (IP), so we had a lot in common. Since the idea of this in-country training at a level that I thought was necessary to overcome the basic O-2A training deficiencies, LTC MacPhearson essentially gave me the lead. He said he would supply me with whatever resources I needed to get the job done. This was great knowing that I had the full support of the 504th Group, which would be needed in order to develop an effective in-country training program. The 504th Group also moved some additional aircraft and maintenance personnel to Nha Trang to support this new mission. These assets were assigned to the 21st TASS since the FOL was part of the 21st TASS's area of operations. The 504th Group remained stationed at Cam Ranh Bay. I developed a syllabus for the O-2A in-country training, which consisted of 10 sorties with emphasis on developing map reading skills. MRII was quite mountainous and covered with triple canopy jungle. LTC MacPhearson approved my syllabus and said that he would help me fly the training sorties. He even moved from Cam Ranh Bay to Nha Trang to assist me.

As I moved south, my living conditions vastly improved. The large bases like Cam Ranh Bay, Nha Trang and Phan Rang were very much like stateside Air Force Bases.

During this time period, President Nixon's Vietnamization Program had been progressing, and many of our U.S. forces had stood down and were returning to the States. With their departure, the 504th Group had access to better flight line facilities and aircraft parking areas for our aircraft. These facilities had previously been utilized by the departed fighter squadrons. For example, at Phu Bai our aircraft were parked in sand-filled bunkers located on PSP decking. At Nha Trang, our aircraft were parked in the same type of sand-filled bunkers but were located on a concrete tarmac. Our flight line office facilities were also much nicer and fully air conditioned with running water, restrooms, and continuous electrical power.

My living quarters at Nha Trang were also much better than the ones at Phu Bai. We had the quarters that the fighter squadrons had lived in. These units had electricity all the time, hot and cold running water and indoor latrines. Nha Trang was much like a stateside Air Force Base. It even had an Officers' Club and a Base Exchange (BX) that was always well stocked. The Air Force knows how to live better than the Army.

At Nha Trang and its Officers' Club, we could get a decent meal. At Phu Bai, we FACs did get to eat meals with the Brigade Commander, so we did not have to go to the Army's Mess Hall. This was okay but while I was at Phu Bai, I would use any excuse to divert to Da Nang if it was around mealtime. At Da Nang we could go to the Navy Mess, which had excellent food. The Navy eats well.

It took approximately two weeks to develop the training syllabus and get it approved. I also set up a training flight operations area on the flight line in one of the departed fighter squadron buildings. I was informed that I would receive my first group of new FACs the next week. I now had a week to set up housing and line up transportation for use during the training.

Since the Air Base had most of the support facilities as stateside bases, I contacted the base transportation unit to arrange for transportation to pick up the new FACs when they arrived. This transportation unit was small and they indicated that they may not always have a driver available on short notice. I asked what the best solution for me would be to cover my transportation needs. They suggested that I get certified to drive one of their buses and after that I could check out a bus whenever I needed it for as long as I needed it. This sounded like a good solution.

The bus certification consisted of me taking a checkout ride with one of their drivers. The driver that checked me out was an Airman First Class. The checkout consisted of me driving the bus around Nha Trang Air Base while he observed. He got a kick out of riding in the bus with a Captain driving it. He would have me honk at people he knew while he waived as we drove around the base. After that, he stamped my military driver's license with "48 Passenger Bus." This turned out to be a very good solution, and transportation was never a problem.

My first class consisted of five newly arrived FACs. LTC MacPhearson and I essentially flew two sorties a day as we provided the newly developed in-country training for these new FACs. After completion of this in-country training, these new FACs were assigned to one of the three TASSs of the 504th Group.

I was at Nha Trang for approximately two months providing this new in-country training for new FACs when the 504th Group decided to move this training activity to Phan Rang Air Base. As the U.S. forces were continually being drawn down, the F-100 Fighter Wing stationed at Phan Rang was sent home. Phan Rang AB was even larger than Nha Trang, and now the vacant fighter facilities offered an even better environment for training.

Phan Rang was located approximately 70 miles south of Nha Trang. I was now on my way to Phan Rang. At least I was moving farther south, and with each move my living conditions improved.

# CHAPTER 18

# In-Country Training at Phan Rang

I ARRIVED AT PHAN Rang on the 1st of December 1971. The living conditions once again greatly improved. Phan Rang was even more like a stateside AFB than Nha Trang. We moved into the housing units and flight line facilities that previously housed the F-100 fighter squadrons. The living quarters were like an 18-unit dormitory type facility with a central latrine area. We also had the previous F-100 flightline facilities from which to conduct our flying activities. These were quite much like stateside facilities. In addition, the runway was long and could handle all types of aircraft. We had plenty of room for our aircraft and maintenance facilities. Phan Rang was also located on the coast, and there were fewer mountains immediately around the base.

There were only two negatives about Phan Rang as far as I was concerned. The first was I had to walk by the Army's mortuary each morning on my way to the flight line. This mortuary was the U.S. casualty unit that prepared our fallen soldiers for their final trip home. Walking by this location each day was sobering because I could see the stacks of wooden coffins ready for use. There were always a large number of these coffins, and it was quite obvious that these coffins were being utilized on a daily basis. There was always a different number of stacks, and each stack always had a different number of coffins waiting to be used. It was obvious that use and resupply was occurring daily. Additionally, the smell of formaldehyde was quite strong. It seemed to take an hour or two each day for me to get the smell out of my mind. This was not a good way to start each day as I headed to the flight line to fly.

The second negative about Phan Rang was when I discovered one of the roads on base had been named after one of the F-100 pilots that had been killed. This pilot was one of those names I had seen on the weekly lists as Killed in Action (KIA) during my fighter qualification training at Cannon AFB prior to my departure from the States. This pilot was also my neighbor and one of my fellow IPs at Laughlin AFB. His name was Mike McGovern, and we were in the same flight at Laughlin AFB.

This road was on a small hill near the center of Phan Rang and was used to get to the top of the hill where the Base's radio towers were located. Every time I looked at this hill, it brought back the memory of my friend, neighbor, and fellow flight member.

Before the end of December of 1971, President Nixon's Vietnamization Program was progressing rapidly. We were notified that the 504th Group was going to be deactivated along with the 19th TASS and sent home. Anyone with less than four months remaining on their tour was going home early. Since I still had six months to go, I was not one of those going home. LTC MacPhearson did have less than four months and he was going home. I was going to be assigned to the 21st TASS at Tan Son Nhut Air Base at Saigon in MRIII. I was again moving south. This ended my in-country training program. I did learn later that my training program had been affectionately named FAC-U by recent graduates. It was known as the Forward Air Controllers "University" at Phan Rang.

# CHAPTER 19

# Tan Son Nhut Operations

I ARRIVED AT TAN Son Nhut Air Base in January 1972. The Vietnamization Program was accelerating. U.S. strength levels in January 1972 were down to approximately 140,000 troops. This was down from a peak level of over 560,000 troops. It was also projected to be down to below 70,000 troops by April 1972. The war was being turned over to the South Vietnam's Army of the Republic of Vietnam (ARVN), the Vietnamese Marine Corps (VNMC) and the Vietnamese Air Force (VNAF). We still had Army and Marine advisors with Vietnamese units, but many of our combat ground forces were standing down.

In MRI the 101st Airborne Division had stood down and was sent home. MRI had been turned over to Vietnamese forces. There were still a few of our ally forces present like the South Koreans and our Special Forces (SF) operating in MRII in the Central Highlands north and west of Nha Trang. The South Korean forces were excellent fighters and quite capable of defending their area of responsibility (AR). Our Special Forces (SF) continued reconnaissance missions in their area of responsibility (AR). They were operating out of Ban Me Thout in MRII.

With the departure of the 504th Group and the deactivation of the 19th TASS, the U.S. had only two Tactical Air Support Squadrons (TASSs) remaining in-country. The 20th TASS remained at Da Nang and still provided the tactical air support required for MRI. The 19th TASS was essentially combined with the 21st TASS which was now located at Tan Son Nhut. The 19th TASS had previously been responsible for tactical air support in MRII. By combining the 19th TASS and the 21st TASS assets into one large squadron, the 21st TASS picked up the responsibility for tactical air support operations in MRII, MRIII and approximately one half of Cambodia. This became a quite large squadron with six FOLs located in MRII and MRIII. The only thing that went home was the name of the Squadron.

Upon my arrival at Tan Son Nhut and the 21st TASS, I was not sure how I would be utilized. LTC MacPhearson, who I had worked for and with at Nha Trang and Phan Rang had gone home with the 504th Group. I essentially was an unknown quantity to the 21st TASS. Nevertheless, I still had six months to go on my tour. I reported into the 21st TASS and waited to see what I was going to do the remaining six months of my tour. It was not long before I found out what was in store for me at the 21st TASS.

As I mentioned earlier, the ranks of the FAC corps were quite young and inexperienced. After I reported in, I found out that the 21st TASS Squadron Commander was a LTC from the Strategic Air Command (SAC). This LTC had been a bomber pilot his entire career, and he was not happy being a FAC. The Maintenance Officer was a Major and was going to retire upon rotation back to the States. I then found out that I was the highest-ranking Captain, and, as such, I was going to be the Operations Officer for the 21st TASS. This was quite a surprise for me as this is usually a Lieutenant Colonel's position.

I was now the Operations Officer of the largest Tactical Air Support Squadron in Vietnam. We had combat tactical air support responsibility for MRII, MRIII and one half of Cambodia. We had approximately 80 pilots, 40 aircraft and 235 Maintenance personnel assigned to provide tactical air support for a very large portion of Vietnam and Cambodia. This support was being provided out of six different operating locations in Vietnam. My Squadron Commander told me that he would take care of all of the paperwork, and I was to take care all of the flying operations including briefings to 7th Air Force. In Vietnam, 7th Air Force had responsibility of all flying operations and they were the ones that issued the daily FRAG Order for combat operations.

I also did not see much of the Major Maintenance Officer as he kept a low, and I mean a very low profile. He was only putting in his time until he went home and retired. He never flew missions, but he did do a quite good job of keeping our aircraft flying. Neither he nor the Squadron Commander ever fly any combat missions.

Air Force organizational structure does not allow for standalone squadrons. With the 504th Group's departure, the 20th and 21st TASSs needed to be assigned to Air Force Wings for administrative support. The 20th TASS was assigned to the Wing at Da Nang while the 21st TASS was assigned to the Wing at Tan Son Nhut. The Wing that the 20th TASS was assigned to at Da Nang was a Fighter Wing and therefore had similar missions. The Wing to which the 21st TASS was assigned to at Tan Son Nhut was an Airlift Wing. The 21st TASS was a combat operational squadron assigned to an Airlift Wing that had totally different missions. This turned out to be a quite serious potential problem for me.

Since we were assigned to the Airlift Wing, as TASS Operations, Officer, I reported to the Airlift Wing Operations Officer who was a full Colonel. I was a mid-range Captain who was filling a position normally held by a LTC and was reporting to a full Colonel. Our initial meeting did not go quite well for me. After I introduced myself, we began to discuss FAC operations. It soon became apparent that this Colonel had no concept of what FACs actually did since he had been an airlift pilot his entire career.

He first asked me about how we scheduled our flights and what type of operational control we had over our airborne aircraft. I explained to him that besides our main operating location here at Tan Son Nhut, we also had six FOLs located in MRII and MRIII that we operated out of daily. I explained to him that each location essentially operated autonomously and prepared and flew their own schedules in support of their areas of responsibility. I further explained that each location had a Tactical Operations Center (TOC) that controlled their own flight operations. This seemed to raise a question in his mind.

He then asked me how many aircraft I had airborne at that minute. I replied that I did not know specifically how many rightnow, but there were probably five or six airborne in MRII and

MRIII and two airborne in Cambodia. He then asked me how we kept track of our airborne aircraft. I again explained the operations and function of each of the TOCs and how each TOC interacted with the two remaining Direct Air Support Centers (DASCs). The problem became evident as he tried to relate FAC operations to airlift operations. He truly did not understand what we did as FACs. His frame of reference for flight operations came from his background of airlift flight operations.

Military Airlift Command (MAC) Wings have total control over all their flight operations. This includes weekly schedules, monthly schedules and even three-month schedules for their aircraft and crews. These schedules consist of assigned crews to specific aircraft for specific missions with specific destinations and return times and dates. He then informed me that since we were now assigned to their Wing, I was going to have to provide him with a weekly schedule, a monthly schedule and even a three-month schedule for all of our flight operations. In addition, he also informed me that I would have to meet these schedules within 3 percent for weekly schedules, 5 percent for monthly schedules and 10 percent for the three-month schedules. He said that I would have to explain any deviations from these percentages at each of the weekly scheduling meetings that I would now have to attend. He also informed me that the proper uniform for these meetings was the 1505 uniform. I was not sure I even had a 1505 uniform with me. I had lived in a flight suit for the last six months.

I knew right then that he probably would not like the flight suit that I had converted into short-sleeved flight suit. When flying O-2As, we did not consider that our greatest concern was protecting our arms from a possible fire. We definitely had other greater concerns and the O-2A did not have air conditioning except by opening the windows.

I respectively tried to explain to him that I truly could not provide any type of daily schedule let alone a weekly, monthly and a three-month schedule. We flew when required to areas where "Charlie" was acting up, and there was no way for me to know our enemy's schedule. I said that the enemy essentially controls our schedule. I did explain that I could provide a daily schedule for the pre-planned air strikes that we knew about when we got the daily

FRAG Order from 7th Air Force each night at 2200 hours (10:00 P.M.). My explanation did not go over well with the Colonel. I could tell that since I was only a Captain and he was a full Colonel, this was going to be a real problem for me.

As an attempt to try and educate him on FAC operations and provide him with more information, I suggested to him that I would be happy to take him around to our FOLs to show him our daily operations. I hoped that this would explain why what he was asking for was going to be extremely difficult for FAC operations. He agreed and we set a date for the following week for him to come to the Squadron for his orientation tour into the FAC world of combat flight operations.

A good thing happened, at least for me, before this scheduled day arrived. "Charlie" decided to launch a rocket attack against Tan Son Nhut AB around 1000 hours (10:00 A.M.) one morning. I immediately launched a FAC to address the situation. My FAC took off and was able to locate the area from which the rockets were being launched. The FAC requested air support from the DASC and two A-37s were scrambled out of Bien Hoa, which was located approximately 20 miles from Saigon. They arrived, and the FAC put them in on the target to address the problem. The air strike silenced the rockets and the FAC returned to Tan Son Nhut.

It was fortunate that this happened since this was not a scheduled flight, and the air strike was visible from Tan Son Nhut AB. The Colonel and other members of the Airlift Wing got a chance firsthand to observe what FACs do. Tell me where "Charlie" will act up, and I will tell you where I will be. This incident definitely helped me explain to the Colonel why I could not meet his rigid scheduling requirements.

The Colonel arrived early on the morning of our scheduled flight to visit our FOLs. The weather was overcast at about 1500 feet with the cloud deck being about 500 to 1000 feet thick. I got all of his information and we went over to Base Operations to file our flight plan. We normally did not go to Base Operations since we would file our FAC flight plans from our TOC. Nevertheless, since this was a full Colonel and this was his Wing, I was trying to comply with what he was used to when flying.

I briefed him that we would first go to Ban Me Thuot FOL, after that Pleiku, followed by Nha Trang, Bien Hoa and then return to Tan Son Nhut. I figured we could visit four out of the six FOLs on this flight. I chose Ban Me Thuot and Pleiku as the first two FOLs to visit since the Army's Special Forces (SF) were operating out of Ben Me Thuot, and the FACs at Pleiku were supporting MRII operations which at times still had active combat operations on going. These two FOLs were in the Central Highlands, which was known not to be friendly.

My first mistake was trying to comply with his expectations with a filed flight plan at Base Operations. I filed a visual flight plan (VFR) instead of an instrument flight plan (IFR). All airlift flight plans are IFR only. He asked me why I did not file an IFR flight plan since we had an overcast cloud deck. I responded tactfully and said VFR is what we should use because Ban Me Thuot did not have a control tower or any active radar control capability. In addition, many of the navigational aids in that area were unreliable.

Ban Me Thuot was not a major base and only had a relatively short asphalt runway with no control tower. It also only had PSP taxiways that went inside a large bunker area where the SFs lived. I conveyed to him that essentially "Charlie" owned the runway at night, since it was outside the bunker area, and we owned it during the day. In addition, the TACAN Navigational Aid was located outside the bunker area close to the runway and was not always reliable.

He then asked me how we were going to get to Ban Me Thuot. I said that we would takeoff, climb-out above the clouds and head north. Ban Me Thuot was located approximately 60 miles north and approximately 20 miles from the Cambodian border. The Army's SFs were operating out of Ban Me Thuot and still were quite active.

FAC missions in support of these SFs consisted of FACs accompanying the insertion missions for Long Range Reconnaissance Patrols (LRRPs) in areas like Cambodia and Laos. Each LRRP team would consist of five to seven SF members who would be inserted into areas of interest for intelligence gathering purposes. The insertion team usually consisted of two Huey helicopters carrying the team along with an O-2A FAC. The Huey's would fly to the insertion point and allow the SFs members to deplane from a hover without actually landing. After the team was inserted, the Huey's and the FAC would move away from the insertion point and loiter for at least an hour. If the insertion was clean and we did not receive any radio transmissions from the team that the insertion had been compromised, we would return to base.

After five or six days, we had a schedule time and location to extract the team. The pickup point was always different from the insertion point. On the missions that I flew in support of pickup, we never got all of the team members back. These SFs members were a totally different breed. During any briefings with these members, I learned that you do not look at them or talk to them unless they started the conversation. These team members only trusted each other and did not like to engage with outsiders.

The missions that these SF members went on were quite dangerous. These individuals essentially were reduced to a basic survival level while doing their duty. These soldiers were truly no longer the normal Army soldier but were killing machines for their own survival while in the field. Some of these SF members did have necklaces of ears that they had cutoff of enemy soldiers that they had encountered. One even had a human skull on his dresser in his room with a candle on top. Many of these SF soldiers became homeless upon return to the States because they could not assimilate back into normal society.

For those of you who understand Maslow's hierarchy of needs, these soldiers had regressed from the higher levels, like self-actualization, down Maslow's pyramid of needs to the basic level of survival. Upon their return to the States, many were not able to recover from their war experiences. War completely changed many of these soldiers.

After we had been flying for about 45 minutes, the Colonel asked me if I knew where we were. The cloud deck was still essentially overcast with only a few breaks in it. I assured him I knew where we were, and I told him that if the clouds were not present, he would be able to see a lake out of his side window. As luck would have it for me, a small break in the clouds was present on his side and sure enough he saw the lake.

Since Ban Me Thuot was a small FOL, the runway was short and not very wide. There also was no control tower present to control aircraft traffic. Ban Me Thuot truly did not have much aircraft traffic except for the SFs helicopters and us FACs. In fact, the Army helicopters usually landed inside the bunker.

When I told the Colonel that Ban Me Thuot did not have a control tower, he asked how we were going to get below the cloud deck to land. Ban Me Thuot, was located in a valley surrounded by mountains. I told him that when we arrived overhead, we would let down in circles until we broke out below the clouds. We would then make a low pass over the runway to clear it and after that land. I then contacted our TOC at Ban Me Thuot and informed them when we would get there. The Colonel seemed to be a little uneasy about how we were going to get down and land.

When we arrived over Ban Me Thuot, the mountains were sticking up through the clouds. I began our descent down into what looked like a bowl of clouds. When we broke out under the clouds right over the runway, I could not tell whether he thought I was lucky or if I actually knew what I was doing. To attempt to try and smooth over our potential issues, I asked him if he would like to try his hand at the landing. He replied, "Yes." I then suggested that we make a low pass over the runway to clear it of anything that should not be there. I told him that since there was no control tower we were on our own.

As we were making the low pass over the runway to clear it, we discovered that an A-1E aircraft had crashed the night before on landing. The A-1E was lying on the overrun at the end of the runway upside down. The Colonel then decided that maybe I should make the landing. We landed over the upside-down A-1E aircraft and taxied to the end of the short runway. There were only two taxiway exits from the runway with one being located at each end. These taxiways consisted of PSP panels that led into the SFs camp which was located behind six-foot dirt bunker walls that surrounded the base. I asked the Colonel if he would watch his side wing tip for me as I watched my wing tip since there was only about a foot of clearance between the bunker walls and our aircraft as we taxied in. As we taxied in, I suggested to the Colonel that he not talk to any of the SFs as they truly did not trust anyone that they did not know. I told him I would show him around our TOC where we could talk to our folks and after that we would be on our way.

As we got out of the aircraft, the Colonel did see some of the SFs members walking nearby. These soldiers all had beards, long hair and uniforms that once were regulation uniforms. Many modifications had been made to these uniforms and many different combinations were in evidence by these individuals. Upon seeing these individuals, the Colonel seemed to increase his pace while we walked to our TOC.

Once inside the TOC, the FACs at this location began to brief the Colonel on their FAC operations. The Colonel seemed to be uneasy about being there at Ban Me Thuot and truly did not seem to be listening. I asked if he would like to see the living quarters of our FACs. He responded with a "No, that's all right," and suggested that we proceed on with the orientation tour. We returned to the aircraft at a rapid pace, started up and taxied out to the runway for takeoff. Since there was no control tower, we taxied out and took off. The Colonel did not say a word as we left Ban Me Thuot. I could tell that he was happy to be out of there.

It was still overcast when we took off so I started circling the field as we climbed-out to avoid the mountains. After we broke out on top of the clouds, we headed north toward Pleiku and the Central Highlands of Vietnam. Pleiku was located in an area that was still somewhat of a "hot" area. Approximately 20 years earlier, the North had infiltrated the area with North Vietnam sympathizers to marry the local inhabitants and assimilate into the population. This had been very effective, and the Viet Cong (VC) were very strong and active in this area. Pleiku was a much larger base and did have a control tower and operated much like a normal Air Force Base. The weather was also much better around Pleiku so our VFR arrival was no problem. We contacted the tower for landing instructions and proceeded to the traffic pattern for landing.

Pleiku had a normal length runway that could handle any size aircraft. They also had concrete taxiways and large covered aircraft parking areas for all types of aircraft. As we turned off the runway onto one of the many taxiways, we noticed that a C-141 was off of the taxiway and stuck in the muddy area between two taxiways. Ground control informed us to use caution to avoid the C-141. Seeing one of his types of airlift aircraft stuck in the mud got the Colonel's attention. We did not know why this aircraft was located where it was, but it was obvious that no one at the present time was working on getting the aircraft out of the mud. I was sure that there was some story behind this aircraft, but we did not know what it was. This seemed to truly get the Colonel's attention.

We taxied into our FAC parking area and shut down. We were met by a couple of the FACs that were assigned at this location who took us into the TOC. The Colonel listened as the FACs described their operations in support of the combat operations being conducted in their area of responsibility. We were supporting ARVN troops as well as some South Korean troops that were still present in MRII. After the briefing, the Colonel was ready to leave.

After we got airborne, I informed the Colonel that we would now visit the Nha Trang FOL. He said that he did not think that was necessary and that he would prefer to return to Tan Son Nhut. I kind of had the feeling that he had seen enough and did not want to know any more about FACs.

After takeoff and his comments, instead of heading east toward Nha Trang, I turned south for return to Tan Son Nhut. The weather had greatly improved, and we were flying above a scattered deck of clouds. We could easily see the tops of the mountains that we were flying over. The Colonel was looking out his side of the aircraft at these mountains when he asked me what the small indentations were along the tops of the mountains. I rolled the aircraft over to his side to see what he was looking at. What he was seeing were 50-caliber gun pits that the VC had constructed along the tops of these mountains. I told him not to worry as there were no guns currently present in these gun pits.

I then relayed to him how the North had infiltrated this part of South Vietnam with Northern sympathizers and that this was not a very friendly area. I told him that this was an area where we had lost a FAC, and when we found his body, his Geneva Convention card had been nailed to his forehead. He did not say anymore the rest of the way back to Tan Son Nhut.

Upon landing at Tan Son Nhut, we got out of the aircraft. The Colonel got out and thanked me for the tour. He then paused and said, "I think I have a better perspective of what your FACs do now. I also saw how you addressed the rocket attack last week, and now I can see you truly do not have a lot to say about your schedule. He then informed me that even though we were assigned to the Airlift Wing, I should do whatever I needed to do to complete our mission. He also said that I did not have to submit any schedules to him, nor did I have to attend his weekly scheduling meetings. He also said that he would provide any support that I needed to do our job. This was very good news to me since I did not have any 1505 uniforms, and I knew I could not meet his requirements for scheduling.

For the next couple of months things were generally quiet as far as the war was going in South Vietnam. Things in Cambodia, nevertheless, were not as quiet. There seemed to be an increase in enemy activity in Cambodia. We were providing tactical air support to the Cambodians who were fighting the Communist Khmer Rouge.

Our area of responsibility in Cambodia extended from the South Vietnamese border west into Cambodia to the Mekong River to Phnon Penh, after that north along the Mekong River to the Laos border. We flew sorties that provided tactical air support for Cambodian "troops-in-contact" (TIC) as well as reconnaissance flights looking for "Charlie" and/or truck traffic. We also provided combat patrol missions in support of the U.S. Navy who would transport supplies to Phnon Penh via the Mekong River. If any of these river convoys were attacked, we would call in air strikes to engage the enemy. Getting clearance to expend ordinance in Cambodia was even harder than in South Vietnam.

I do remember a couple of my missions in Cambodia that stood out from the others. On one mission when I was flying a reconnaissance mission along the Mekong River, I started taking 23-mm fire from a Pagoda. This Pagoda was located in a small village that consisted of 30 to 40 structures. I reported the 23-mm activity to our TOC and proceeded away from the area. I gave the TOC the coordinates of the 23-mm because I knew I could not get clearance because of our Rules of Engagement (ROE).

Our ROE stated that we could not engage the enemy in any Pagoda nor could we strike any village with 25 or more structures. Since this target failed both of these rules, I decided to get out of the area. 23-mm guns are not fun targets to attack since they can fire up to 700 rounds per minute. In addition, we flew well within their engagement envelope. Quad 23-mm (four barrels) guns were a very serious problem for us. Thankfully, this one was only a single barrel.

I was leaving the area to look elsewhere when our TOC contacted me and asked if I wanted clearance for the target. I again reminded them that the gun was firing out of a Pagoda window that was located in a village that had more than 25 structures. I am sure that "Charlie" also knew our ROE and, as such, he felt quite safe firing at me. The TOC responded with a "Yes," that they understood, but do I want clearance?" I responded by saying, "If you can get clearance, then yes, by all means, I want it." Five minutes later I received clearance and was told fighters were on the way.

When the fighters checked in, I briefed them on the target and told them to watch their tails since this was an active 23-mm site. After 10 minutes, the Pagoda was destroyed along with several nearby structures. There also were secondary explosions going off as we must have hit an ammo storage facility as well. That was a very good mission, but I never understood how I got clearance and I never asked.

A second mission in Cambodia that stood out occurred when I was again on a reconnaissance mission. I spotted a convoy of trucks on a road that was heading into the Chup Rubber Plantation. The Chup Rubber Plantation was a very large plantation that supplied raw materials to the Michelin Tire Company. The Chup Rubber Plantation was also one of those targets that went against our ROE.

The truck convoy entered the Plantation from the north and after that disappeared. This peaked my interest. I made a number of passes over the Plantation attempting to locate the truck convoy. From the air and the view that I had, I knew there was no way that they could conceal all of those trucks above ground so fast. I had to assume the trucks had entered a large underground bunker since they were nowhere to be seen. Since the Chup Plantation was on the "no strike" list, I called in what I had observed to our TOC and proceeded on my way.

About 10 minutes later the TOC contacted me and asked me if I wanted clearance. I responded with "You know that this is the Chup Plantation, don't you?" They confirmed the location as the Plantation and again asked if I wanted clearance. I replied, "If you can get clearance, then by all means I want it." I knew approximately where the trucks had disappeared after they entered the Plantation. Approximately 20 minutes later the TOC indicated that we had clearance. I then requested any air available.

A set of F-4s were the first to check-in. I briefed them on the target and told them we were going after an underground bunker that was a truck park. I marked the target area and requested that each fighter lay down a string of their 500-lb bombs in the target area with a separation of 100 to 200 feet between the strings. After the F-4s had expended their ordinance, I made a pass over the target area to assess the damage. We were right on target. I noted large tree logs, as much as 24 inches in diameter, that were splintered and sticking out of the ground. These logs were obviously the roof support structure of the underground bunker. I immediately contacted our TOC and requested more fighters.

I put in two more sets of F-4s that afternoon. By the second set of fighters, we had done enough damage to the bunker roof structure that the bombs from the last set of fighters actually penetrated the bunker itself. There were numerous secondary explosions going off with debris being thrown up to 2000 feet in the air. In addition, there were large fires burning and there was smoke everywhere. We had hit a major stockpile of ammunition in an underground storage bunker in this Plantation. I am sure that "Charlie" felt his supplies would be safe in this underground bunker since it was the Chup Plantation. I am also sure that he knew our ROE or he would not have been so bold as to be moving these supplies in broad daylight.

Later that day an RF-4C reconnaissance bird made a pass over the target to assess the damage. They reported that secondaries were still going off. I am not sure who actually gave us clearance to strike this target, but we sure did hit the jackpot on that day. I never again got clearance to hit the Chup Plantation. I also found out later that our government reimbursed the owners of the Plantation $300 for each tree that was destroyed. To me it was worth it.

It wasn't until later that I was able to put two and two together as to why I received clearance on those two targets. I determined that our intelligence had information that something big was about to happen.

It was also during this time period that I had the experience of God speaking to me. I was scheduled for a day reconnaissance mission, and as I was taxing out to the end of the runway for takeoff. I felt a very strong, overwhelming feeling come over me that something was wrong. There also was a distinct voice in my head saying, "Don't fly Today." I taxied a little further and the voice got a little louder and more stern and said "Don't Fly Today." I continued to taxi and as I approached the runway, the voice said again but this time in a quite loud and stern voice "DON'T FLY TODAY." I have never experienced that type of feeling before on any other of my combat missions nor before or after on any of my flights during my twenty-five years in the Air Force. I aborted the takeoff and taxied back to the parking area. I know that God had spoken to me, and he did not want me to fly that day. When God spoke to me, I listened! Since I was the Operations Officer, I did not have to explain why I did not fly to anyone. I trusted God and that is when I knew God was guiding my life.

# CHAPTER 20

# The Spring Offensive of 1972

ON MARCH 30, 1972, everything changed as far as the war was concerned. North Vietnam launched a major offensive into MRI. Troops came across the DMZ south and from the west out of Laos into Quang Tri Province. They were quite aggressive and caught the South Vietnamese forces by surprise. Since MRI was under the control of the South Vietnamese forces, they redeployed some of their units from MRIII to assist MRI. The 20th TASS was still present at Da Nang and was able to provide tactical air support from the U.S. units that were still available. The U.S. Navy still had assets off of the coast, and the Air Force still had fighter support based in Thailand. The Army and Marines also still had advisors with the ARVN units so they still had access to U.S. tactical air support.

It was not long before we found out that the North had planned a three-prong attack on the South. A second front was opened up by the North in MRII west of Pleiku near Kontum in the Central Highlands. North Vietnamese troops entered South Vietnam from the southern Laos/northern Cambodian border area through the An Kay Pass. These troops were heading east and seemed to have as a goal of cutting the South in half. If they could get to the coast, they would have access to a seaport from which they could resupply their forces by sea and not have to depend so heavily on truck traffic and the Ho Chi Minh Trail system of resupply. If successful, this would be a tremendous boost to their combat capability. In MRIII, the third prong of their offensive was initiated out of Cambodia near the South Vietnamese village of Loc Ninh. Loc Ninh was located about 100 miles north of Saigon and had a major road that ran south through An Loc to Saigon.

It soon became obvious that this three-prong attack was a major offensive by the North to take over South Vietnam. I think the North felt that since the U.S. was withdrawing its forces rapidly, they felt they could now achieve their goal of taking over the South and embarrass the United States as we withdrew. This three-prong attack put a significant strain on the South Vietnamese forces as well as the remaining U.S forces. Two of the three-prongs of the attack by the North were in MRII and MRIII–my areas of responsibilities for tactical air support. My life as well as all of the FACs in MRII and MRIII changed significantly and in a very short period of time. All of the FOLs in MRII and MRIII became very active with a "hot" war again.

It was a very good thing that each FOL operated essentially autonomously. It would have been virtually impossible to control all of combat operations if operational control had been attempted by a central location. It was also good that the Airlift Wing decided to leave us alone and provide any support that we needed. They now truly understood why we could not meet their scheduling requirements of a centralized command organizational structure. We were once again in a "hot" war now on three specific fronts with the ground combat load being carried by South Vietnamese forces.

A number of other things became clear by this offensive outbreak by the North. It was now evident that the missions I had flown in Cambodia against the 23-mm and the Chup Plantation were a prelude to this offensive. This was evidenced by the large number of secondary explosions that occurred with the targeting of these two locations. Both targets were obviously large ammunition stockpiles that were being prepared for this offensive. In addition, the mission where I encountered the 23-mm in Cambodia also suggested that they were preparing for something big. We normally only encountered 23-mm when something big was going to happen and normally not this far south. It was all becoming quite clear as to the intent of the North.

The FACs in MRII had also been reporting 37-mm guns present in that region. This was the farthest south that we had encountered that size of antiaircraft guns. We also were beginning to encounter SA-7 shoulder launched missiles. With all of this increase in activity, it should have been obvious that the North was indeed planning something big. The amount and size of these weapons suggested that regular NVA forces were also present. The VC had not tried to engage us with anything larger than 50-caliber guns other than an occasional 23-mm gun. The presence of 37-mm and SA-7 missiles represented a whole new level of battle for us. The 37-mm did not bother us too much since these rounds were airburst rounds that usually went off well above our altitude, and they could only fire seven rounds per minute. The SA-7, however, did represent a significant threat to FACs.

Our flight operations placed us well within the SA-7 missiles envelope of engagement. In addition, the O-2A did not have any defensive system to defend itself against this shoulder launched missiles. These missiles had an infrared guidance system and once launched, they did not need radar or any other type of ground control. Once launched, the infrared sensor would start searching for a heat source as its target. The O-2A, with its two engines that were heat sources, were prime targets for SA-7 missiles.

After some losses to SA-7 missiles, we did however, develop a defensive maneuver against these missiles, if we were lucky enough to see the missile launched. Of the two engines on the O-2A, the rear engine ran about 200 to 300 degrees hotter than the front engine due to less cooling airflow. The enemy also usually launched the missile after we passed by them. This, unfortunately, exposed the highest heat source for the missile to lock on to.

The defensive maneuver that we developed was almost a suicide maneuver. If we did see the missile launch, we would wait until it locked on to one of our engines. We could tell when it locked on because as it left the launcher, it had a corkscrew flight path of smoke while in the search mode. When it locked on, it would kick in toward us, and the smoke trail became steady as the missile's sensor tracked its target.

The maneuver that we would perform was to watch the corkscrew flight path until it locked on to one of our engines. This was evidenced by a sharp kick into us followed by a steady smoke stream trail heading toward us. We would then make a hard break into the missile like we were playing chicken with the missile. By doing this maneuver, we hoped that the missile had locked onto the rear engine since it ran hotter than the front engine. By breaking into the missile, we were shielding the rear engine from the missile with the front engine and the cockpit. If the missile had locked onto the rear engine, it would break lock and start searching again as evidenced by a corkscrew flight path.

During this maneuver, it was helpful if the weather conditions consisted of a scattered to broken deck of clouds above us. Many times the missile's sensor would lock on to the heat of the sun coming through a break in the clouds. The missile seemed to like the sun's heat more than the heat of our front engine. This maneuver worked about 90 percent of the time, if we were lucky enough to see the missile launched. At our altitude, the time of flight for these missiles before contact was five to seven seconds.

One of my young Lieutenant FACs encountered a quad launch of SA-7 missiles on one of his missions. It was the first time that we had encountered a quad launch of SA-7 missiles. He reported that the launch of the four missiles was almost simultaneously, which was lucky for him. In addition, all four missiles seemed to lock on at nearly the same time. When they did lock on, the Lieutenant (LT) performed the hard break into all four of the missiles. Fortunately for him, all four missiles had locked onto the rear engine and all four missiles did break lock. Two of the missiles passed over the top of his right wing while the other two missiles passed under his right wing. He immediately returned to base having been totally shaken by this near miss. As he was telling me about his encounter, he was still quite white and his hands were shaking. God was definitely watching over him that day. After a couple of days off from flying, he once again rejoined the battle.

With the North initiating this three-prong attack and two of the three prongs located in MRII and MRIII, the 21st TASS became extremely busy. We had to provide 24-hour airborne coverage in both these regions and still support Cambodia. Our FACs began flying two or three missions every day in support of combat operations to stop the North. This was a very intensive time period for all of us.

As Squadron Operations Officer, I decided that I would fly two missions each night and then run the Squadron during the day to keep up the required support to the FOLs. With the tremendous increase in flying time, the time between required periodic inspections of the aircraft was greatly reduced. This caused me to have to move aircraft around to keep safe aircraft in the field for use. I also had to move FACs around to different FOLs as required to meet the threat.

For the next 30 days (the month of April 1972), I would sleep two hours between 1800 – 2000 hours (6:00 P.M. to 8:00 P.M.) andthen fly two-night missions. Upon completion of these two missions, I would sleep between 0600-0800 hours (6:00 A.M. to 8:00 A.M.) and then run the Squadron until 1800 hours when I would repeat the cycle. During this time period, I lost about 25 pounds because I did not truly have time to eat properly. I ate on the run whatever I could get a hold of that was food and fast. I do not recommend this as a good diet plan.

The offensive in MRIII started on the 6th of April 1972 at the village of Loc Ninh. Loc Ninh is located approximately 100 miles north of Saigon near the Cambodian border. I was the FAC on station when the offensive started. An American Marine advisor with the ARVN unit at this location requested immediate air support when a large North Vietnamese Army unit attacked their fire base. There was no doubt that this was going to be a major offensive. The ground action reminded me totally of my mission is support of FB Sarge when I was assigned to Phu Bai in support of the 101st Airborne. The only difference this time was that these were ARVN forces and a Marine Major advisor on the ground instead of the 101st Airborne. I flew two missions in support of this FB that night, which essentially started my intense schedule for the next 30 days.

The next day, the ARVN unit at this FB had to retreat to An Loc which was approximately 20 miles south of Loc Ninh. The FB at An Loc was larger than the one at Loc Ninh so the ARVN decided it was better to defend at An Loc. The battle for An Loc was fierce and brutal. During the day I had three airborne FACs on station to support this FB. At night I had two airborne FACs on station to provide support.

Our battle plan for defense of this FB consisted of the three FACs during the day. One FAC would act as a gatekeeper and keep track of the fighter aircraft as they checked in. The fighters were stacked in a holding pattern until one of the other two FACs were ready to put them in. This allowed us to have constant air support during the day. At night we could handle the air strikes with two FACs since the number of air strikes was reduced but still more than one FAC could handle. I used my more experienced FACs to direct the fighters and the younger less experienced FACs to act as the gatekeepers. This worked pretty well.

The battle for An Loc went on for approximately 10 days. The North was taking heavy losses but were resupplying and reinforcing their units daily. For us to resupply the ARVN forces at An Loc, we used C-130 aircraft, which would air drop supplies for the FB. This was a quite dangerous mission for our C-130 aircrews as they had to fly quite low and airdrop the supplies next to the FB. We lost one C-130 during one of these resupply missions. We could not get any helicopters into the area to rescue the crew due to ground fire. The ARVN from the FB, however, were able to recover the C-130 aircrew. It took three days before we could get the crew back to Saigon. The pilot of the C-130 had been a fellow IP with me at Laughlin AFB, and I had seen him the day before.

With the schedule that we were flying we were rapidly flying our aircraft out of flying time between inspections. Our maintenance personnel worked their tails off trying to keep our aircraft flying. We were forced to start waving these required inspections because we could not go without the aircraft. We also were not able to comply with the Air Force's crew rest policy for our pilots.

I remember one night when I returned from my first mission of the night, I called into our maintenance to let them know I needed to return to An Loc as soon as possible. I was informed that we did not have any serviceable aircraft ready to fly. I inquired what was wrong with each of the aircraft that had been grounded from flight. Most of the aircraft had engine problems which required repair or engine replacement. I asked if any of the grounded aircraft had two good engines. The response was "Yes, but that aircraft was missing the attitude indicator." This instrument is the primary flight control instrument for all aircraft. The attitude indicator is the flight instrument that pilots use to maintain level flight in weather or at night. I told them to prepare that aircraft for flight with fuel and rockets because I did not have a choice. I had to return to An Loc since the lives of those on the ground were depending upon me to give them the tactical air support they needed. I knew the weather was good between Tan Son Nhut and An Loc and I would not be looking inside the aircraft to maintain control.

This probably was not the smartest decision on my part but the situation was critical for those on the ground. I also was not going to ask anyone else to go. In fact, I would have gotten after anyone who would do such a thing. In war, however, you do whatever is required to save lives.

It was also during this time period that I developed my opinion about why I felt the Cambodians were better soldiers and cared more for their country than many of the South Vietnamese did. A number of times I had to load my own rockets when our munitions personnel were overloaded by the schedule and had not been able to get all of the aircraft ready for flight. We had hired some South Vietnamese to help with rocket building and loading of rockets onto the aircraft. This helped considerably until the monsoon season started.

When the monsoon season started and we would have a shower, many of these Vietnamese workers would not show up to build and load rockets. They were more interested in gathering "rice bugs" from drainage ditches when it rained enough to have water running in these ditches, which ran alongside the roads on Base. Evidently these "rice bugs" were a delicacy and I was surprised at the number of Vietnamese that would be collecting these bugs after each rain. They would bite the heads off of these bugs and then suck out the insides. It appeared these bugs meant more than saving their country.

I was also the FAC on station when An Loc fell. I was working with the Marine Advisor when he informed me that the NVA had breached their perimeter. He said that "Charlie" was coming through the wire and that I should hit his position. I asked him if he was sure. He responded with a "Yes, that he would not be able to tell if the ordinance hitting him was ours or theirs." He then said, " hit my position." His next radio transmission truly hit home. He said, "My wife is in Bangkok. Tell her I love her." The radio then went completely silent. That was one of the worst moments during my entire time in Vietnam. I will never forget that feeling of helplessness that I felt at that moment.

With the radio being silent, I called our TOC to see what they wanted with the remaining air strikes that were waiting. They responded by saying hit the FB and after that take out the northern six blocks of An Loc. They said there are no friendlies left in An Loc. I completed the mission and returned to Tan Son Nhut. I was devastated by the events of the night. I knew the situation was tough on the ground, but this mission truly brought it home.

For the next two weeks our efforts were to try and stop the advance by the North down the highway from An Loc to Saigon. They were now only about 70 miles north of Saigon. The South Vietnamese deployed their troops along this highway north of Saigon in their attempt to stop the advance. We continued FAC coverage day and night to assist in their efforts.

It was three days after my last contact with the Marine Major at An Loc when I got the best news that I could have received. I was at my desk working on the schedule when one of my Sergeants came into me and said, "Someone is looking for Chico 10." Chico 10 was my call sign when I was flying out of Tan Son Nhut. I responded with "Bring him in." In walked a Marine Major wearing a poncho and covered with red dirt and sporting a 10-day growth of beard. He had red hair, which almost matched the red dirt.

When he saw me, he asked, "Are you Chico 10?" I responded with a "Yes." He rushed over to me and said, "I want to shake your hand. I am the guy whose life you saved at An Loc." He said when the air strikes began to hit his position, he was able to escape to the south from the FB. He indicated that it had taken him three days to "escape and evade Charlie" and he wanted to thank me for saving his life. He grabbed my hand and began shaking it with great enthusiasm. I smiled at him and said, "You can tell your wife that you love her." He smiled back and at that moment a bond forever had been established between him and me. I do not remember his name, but I am sure he remembers me as well as I remember him, without names — just brothers in combat.

Before he left, I asked him if I could get him anything. He responded with, "I could use some dry socks." I had him sit down and told him I would be right back. An Army supply unit was next door to our flight operations building. I hurried over to the supply warehouse and told the Army Sergeant what I needed and for whom. Without any hesitation, he gave me a dozen new pair of socks for the Major. I returned to my office and gave the Major the socks. His face lit up like a little child getting a Christmas present from Santa Claus. He took the socks, stuffed them in his backpack, shook my hand again and thanked me once more and left. He said that he was on his way to Bien Hoa where he would be assigned to another ARVN unit and go back up the road to stop "Charlie." I never met a Marine I did not like. They are awesome!

My night missions over An Loc left me ingrained with an extremely strong dislike of two quite normal things. I now do not like any fireworks displays even on the 4th of July. Fireworks remind me of 23-and 37-mm guns. My second dislike is of those large search lights that car dealerships and others place outside their businesses to draw attention to their location. "Charlie" was using these types of search lights trying to find us over An Loc. We flew our night missions without exterior lights so they would try to locate us with these search lights. They knew if they could shoot down the FAC, air strikes were less likely to cause them problems, especially at night. To this day whenever I see fireworks or those search lights, I am again immediately in an O-2A over An Loc.

It was also during this time period that I encountered a very difficult situation with one of the new FACs who had arrived. The individual was a young Captain who had been flying C-141s after completion of pilot training, and this was his first assignment after that. After he processed into the Squadron, I was talking with him and explaining how we operated and what would be required of him. He was very quiet and did not respond much in any way. I told him I would schedule him with an experienced pilot to check him out. I told him we would start his checkout the next day. He then asked me if there was any way that I could get him assigned to Korea. We did have O-2A FACs in Korea, but he had been assigned to Vietnam. In fact, some of the Korean FACs had been sent TDY (Temporary Duty) to Vietnam when we were short on FACs. His question raised a significant concern since I had never been asked that question before. I told him that I did not have the authority or capability to get him transferred to Korea. I told him to check the schedule in the morning as I would have him scheduled to fly.

The next morning this Captain came into Operations to check the schedule. I had scheduled him for a flight in support of Cambodia. I did not want to start him out on a mission in a real "hot" area. After looking at the schedule, he asked if he could talk to me. I, of course, agreed and took him over to my desk. He sat down but did not say anything. After a couple of minutes of silence with him looking down at the floor, I asked him what he wanted to talk about. In a quite soft voice, without looking up from the floor, he responded by saying, "I can't fly." This, of course, did not go over well with me since all of my other pilots were flying their tails off. I responded by saying, "You have to fly." I then went on to say that I could not have him around the Squadron and not be flying missions. I then told him that if he was refusing to fly combat, I would have to ship him back to the States with a recommendation for a Court Martial. I then asked him why he could not fly. He responded with, "I can't." I told him we were not leaving my office until he told me why he could not fly combat. He kept saying, "I can't fly combat missions."

After we sat there for about 10 to 15 minutes with him not saying anything, I asked him if he could not tell me why he couldn't fly combat missions, could he write it down for me? He responded with a "Yes," he thought he could do that. I got him a tablet and pencil and took him over to an empty desk that was not far from me. I then went back to my desk to work and left him alone to write.

I kept watching him from my desk, but I did not see much writing going on. After about an hour and a half, I got up and went over to the desk he was at and asked him if he was finished. He did not look up, but he did respond with a "Yes." He handed me the tablet that I had given him to write on. On the tablet were four words. They were "I can't take life." I took him back over to my desk to try to get more information from him. I asked him to explain what he meant by his written statement. He paused and after that said, "Well, if I put in an air strike which kills the bad guys, I take life. If I don't put in the air strike and the bad guys kill the good guys, after that I am responsible for their loss of life. Either way I would be responsible for the loss of life." I then asked him if this was part of his religious beliefs. He responded with a "Yes." This caused me to pause since I now had a Captain who was refusing to fly combat due to his religious beliefs. This was definitely going to be a problem for me.

I then asked him why he even joined the Air Force in the first place if he felt that way. He paused and then said that he felt he had an obligation to serve his country. He went on to say that he felt that he could do that by flying airlift-type missions like those flown by C-141s. He said he never thought that he would be asked to fly combat missions, which would require him to make decisions on who lived or died. When he got the FAC assignment to O-2As, he asked to go to Korea since he knew that he would not be flying missions where this type of decision would be required. He did not know that some of the Korean O-2A FACs were coming to Vietnam on temporary duty. This caused me to pause because I truly felt he was telling me the truth.

I did not say anything for a while as I was trying to decide what I was going to do. I could not have him around the Squadron and not be flying. I then asked him how long he had been in the Air Force. He responded with a little over four years. I the asked him if he planned to stay in the Air Force for a career. He responded with a "No, he was not going to now, since he now realized that he may be placed in similar situations in the future."

This gave me an idea. I did not have the authority to get him transferred to Korea, but I did have the capability to get him assigned to a non-flying job at the Direct Air Support Center (DASC). At the DASC his job would entail the coordination of getting fighters to FACs and obtaining the required clearances for targets. I asked him if he felt that he could complete that type of duty. He responded with he thought he could do that. He said that he felt by doing this type of job he would not directly be responsible for taking life. I then told him that if he went over to Personnel and requested a "date of separation" (DOS) from the Air Force upon rotation back to the States, I would get him assigned to the DASC. This tour in Vietnam would fulfill his obligation to the Air Force after pilot training. He said that he would do that. The next day the Captain went to Personnel and obtained a DOS upon his return to the States. I then made arrangements to get him assigned to the DASC. The Captain departed and both of our problems had been solved. I had reached a solution that both of us could live with.

The ARVN were now beginning to slow the NVA advance to the south from AnLoc in MRIII. The NVA advance in MRII was, however, still going strong. I had to shift some of our assets north to FOLs in MRII to help counter this advance by the North. I also began flying missions in MRII myself out of Tan Son Nhut. At times, I would fly two or three missions before returning to Saigon.

In MRII the North had broken through An Kay Pass and were heading east toward either Kontrum or Pleiku. The remaining South Korean forces that were still in-country were trying to retake the An Kay Pass. As I have previously noted, the South Korean forces were fierce soldiers, and I truly respected their efforts. I know why North Korea had not tried to take over South Korea after the fighting ceased in Korea back in the 1950s.

I gained this respect on one of these missions in MRII while supporting the South Koreans. The NVA had constructed concrete pill boxes in the mountains around An Kay Pass. The South Koreans were taking heavy losses from these pill boxes as they were attempting to take the Pass back. The South Koreans had requested our tactical air support to assist in neutralizing one of these pill boxes. We had 2000-lb bombs that could neutralize this threat. It was on one of these missions when some South Korean troops had been pinned down by heavy 50-caliber machine gun fire from one of these pill boxes when they requested our support.

I arrived on station and was briefed by the Korean Ground Commander on their situation. I knew it was going to be difficult to provide them the air support that they needed. Their position was such that they were quite close to the target; and, due to the mountains, our fighters would have to overfly their position to attack the target. It was on the second set of fighters that I was directing when an incident occurred that all FACs dreaded—a "short round."

After the fighters checked in with me, I briefed them on the situation. I informed them that the target was a concrete pill box with active 50-caliber machine guns present. I also informed them of the restricted access to the target due to the mountains, which would require overflight of the friendlies (South Koreans).

I asked the fighters if they understood the situation and if they could handle it. They both responded that they understood and that they could comply. I marked the target and asked the South Koreans to pop smoke to identify their position to the fighters. Both fighters acknowledged they had the target location and the friendly location. We proceeded with the attack.

Lead rolled in and after I visually verified his location, I cleared him hot. Both of his 2000-lb bombs landed right on target. As two rolled in and he confirmed the target location and the friendly location, I cleared him hot. My worse fear came to fruition as one of his two 2000-lb bombs released prematurely and exploded directly on the friendly location. I directed both fighters to "safe them up and hold high and dry," while I checked with the South Koreans.

I contacted the Korean Ground Commander to see if the last bomb had been a problem. He responded with a "No" and to "keep dropping bombs." I asked him if he was sure, and he responded with "We will lose more people if you don't get the pill box." We continued the attack and put in a third set of fighters.

We were able to silence the pill box that day, but I know the "short round" had to have impacted the South Koreans. They were, however, happy with the results of the pill box and I never heard a word about the "short round." I gained even more respect for the toughness of the South Korean forces that day.

Over the next two or three days, the NVA were finally able to break completely through the An Kay Pass and head east. It soon became apparent that the North was heading to Kontum and not Pleiku. They must have felt it would be easier to take Kontum instead of Pleiku. The U.S. still had a significant present at Pleiku.

From Kontum they headed east toward Qui Nhon, which was on the east coast and could be a seaport for their resupply. Their order of battle seemed to be to approach Qui Nhon from the west and north. Some of the NVA forces were approaching the coast approximately 40 miles north of Qui Nhon and the small village of Bong Son. This was a small village along Highway 1 which runs along the entire coast of Vietnam. This Highway runs from Saigon north along the coast to the DMZ. At the DMZ, Highway 1 continues along North Vietnam's east coast to Hanoi. Highway 1 is the major highway for both North and South Vietnam.

In MRII, the North was still advancing. This area was not as populated as MRI and MRIII. As noted earlier, the North had infiltrated this area with their sympathizers 20 years earlier. This area had a lot of local support for the North's efforts. In addition, most of our U.S. Army forces in this area had been sent home as part of President Nixon's Vietnamization Program. That left this area more vulnerable than MRI and MRIII. The South Vietnamese forces in this area were less experienced, which resulted in less combat capability being available.

When the U.S. Army Forces in MRII left, an Army Lieutenant Colonel (LTC), who had married a Vietnamese woman, had retired, and was going to remain in-country. His name was John Paul Vann. The Central Intelligence Agency (CIA), hired him to become the MRII Commander. This gave him significant power and resources. I had two missions in MRII that were at the direct request of Mr. John Paul Vann. These two missions were against the NVA, and both were around the Bong Son area along Highway 1.

The first mission that Mr. Vann requested, and I supported was when the NVA was approaching the eastern coast of South Vietnam at the village of Bong Son, which was north of Qui Nhon. The South Vietnamese forces that had been stationed at a FB near Bong Son had abandoned their FB and had fled south. When they left this FB, they did not disable their 105 Howitzers. Instead, they left 12 of these Howitzers in perfect working condition for the NVA troops to acquire and use against the South.

Mr. Vann was able to get the Navy to provide him a fire support mission from the Battleship Missouri. This Battleship had been recommissioned and was stationed with the 7th Fleet off the east coast of Vietnam. Mr. Vann wanted a fire support mission from the Missouri and their 16-inch guns to destroy the abandoned 105 Howitzers at the FB. These 16-inch guns were capable of shooting an approximate 250-lb projectile for distances of 16 to 17 miles.

When I arrived on station, the weather was perfect; and I could clearly see the Missouri off of the coast to the east. Mr. Vann briefed me from his helicopter on the location of the Howitzers and said he wanted them all destroyed before the North got to this FB. I then contacted the Missouri's Fire Control Officer and briefed him on the target. I provided him with the target coordinates, and we proceeded with the fire mission. It was the first and only time that I directed Navy gunfire, and I was impressed!

The Fire Control Officer would give me "time of flight" for each round fired. The "time of flight" for these rounds from the Missouri's location to the target ranged from 55 to 58 seconds. The target was approximately 12 to 13 miles from the Missouri. The Fire Control Officer would call "Round Out," and I would observe a puff of white smoke from one of the 16-inch guns. I would start my clock and then observe the target for impact. I would then make any adjustments on the coordinates, and after that we would repeat the process. After approximately 45 minutes, we were able to destroy all of the Howitzers that had been abandoned. I was truly impressed how far those 16-inch guns could shoot and how accurate they were at that distance. I truly enjoyed working with the Navy on this fire mission.

The second mission I flew in support of Mr. Vann occurred a couple of days later. The NVA had captured the village of Bong Son and were prepared to advance south toward Qui Nhon along Highway 1. When I arrived on station for this mission, Mr. Vann again briefed me from his helicopter on the target. He said that he no longer wanted the center span of the Bong Son River Bridge. This was a major bridge along Highway 1 and certainly would restrict the NVA from moving south with tanks or other heavy artillery weapons. Mr. Vann went on to say that he had F-4 aircraft inbound with laser-guided bombs to destroy the bridge.

I was totally impressed that Mr. Vann had been able to get a battleship for a fire support mission and now he was able to get laser-guided bombs. This was the only time I ever had laser-guided bombs on fighters that checked in with me. Laser-guided bombs were new and quite expensive. Like I said, Mr. Vann had significant pull with someone quite high up to get this type of resources for the support he needed.

When the F-4s checked in with me, they each had four laser-guided 500-lb bombs. I briefed them on the target as the center span of the river bridge along Highway 1 coming out of the village of Bong Son. The target was quite obvious, so I did not have to mark it. I cleared lead in hot as I watched the bridge. Lead's bomb hit the center stripe of the center span and caused significant damage. I cleared two in to hit the remaining bridge structure of the center span. After the two bombs had impacted the target, the center span dropped into the river. Lead then asked where I wanted the remaining bombs. I contacted Mr. Vann to see where he wanted the remaining bombs dropped. He responded by saying, "Drop the rest of the bombs on Bong Son Village. There are no friendlies remaining in Bong Son." The remaining six bombs were dropped on the village.

I was totally amazed at the amount of power Mr. Vann had in prosecuting the war in MRII during this offensive. That was the last time that I had any contact with Mr. Vann. I will say that in my opinion the decision to destroy the river bridge on Highway 1 by Mr. Vann was the critical decision that resulted in the stopping the 1972 Spring Offensive. By stopping the North from gaining a seaport at Qui Nhon, they were not able to effectively resupply their troops. Within a week, the level of hostilities in both MRII and MRIII began to subside. The only sad note was when I learned that Mr. Vann had been killed in a helicopter crash on June 9, 1972, near Kontum. He was a real hero in my opinion.

It was also during this time period when I was flying missions in MRII out of Saigon that I experienced the loss of the rear engine for the second time. I had flown three missions that day in MRII while checking out a new FAC. It was on our fourth mission, and we were returning to Tan Son Nhut at night. We had taken off from Pleuki and had climbed up to an altitude of about 8500 feet and were heading south. As previously noted, MRII consisted primarily of mountains in the Central Highlands. In fact, the highest peak in South Vietnam is Ngoc Linh at 8524 feet and is north of Kontun and Pleuki. In addition, as I have also previously mentioned, the VC did not take American prisoners, especially in MRII.

We had leveled off at our cruising altitude of 8500 feet when we lost the rear engine. As previously noted, with only the front engine one could only maintain approximately 1000 feet of altitude for sustained level flight. This meant that we were coming down to at least 1000 feet. We were at night above a solid cloud deck and above mountains in a region where they did not take American prisoners. This did not leave many options for us to consider. Going back to Pleuki was out of the question due to the height of the mountains. If we bailed out and were able to survive the bailout at night, it was quite well known that the VC in this area did not take prisoners. As previously noted, the last pilot to go down in this area was found with his Geneva Convention card nailed to his forehead.

I explained the situation to this new young Lieutenant (LT) FAC, since I felt he definitely had a say as to what course of action we should pursue. I also told him I felt that we did have enough altitude that we could lose and make it to the coast and Nha Trang. I hoped that our slow descent from 8500 feet to 1000 feet would end up over the water east of Nha Trang. If not, the cloud deck would be full of granite. Being new, the young LT said he would let me make the decision based on my experience. I decided that we would try to make the Nha Trang. We turned southeast and headed toward Nha Trang as we started our slow descent.

God was definitely watching over us that night because after entering the cloud deck, we broke out over the Bay of Nha Trang about two miles east of the end of the runway. Nha Trang was also surrounded by mountains with the base being located in a valley. There are mountains on three sides, and the Bay of Nha Trang is on the east. We leveled off at 1000 feet under the clouds, turned around and landed uneventful at Nha Trang. It took two days to get another engine from Tan Son Nhut to fix the aircraft after which we returned to Saigon.

During my last month in-country (May 1972), I experienced the loss of three FACs. I took these losses quite personally. The first loss was a young Lieutenant (LT) that I had checked out. He was flying in Cambodia when he was shot down. I had completed his checkout the week before. Being the Operations Officer, I had to clean out his room and prepare his belongings to be sent home. It is not a pleasant task to clean out a person's belongings to be sent home.

The next of kin that this LT had listed was his grandmother. In addition to cleaning out his room, I had to write the letter to his grandmother for the Squadron Commander's signature. The letter was to accompany his personal items. I took his loss quite personally since I had finished his checkout. I wondered if there was something that I had missed during his checkout that I should have covered that would have saved his life.

The question was answered approximately two weeks later. The investigation into this loss determined that he was at approximately 700 feet firing his AR15 out the window during a "troops-in-contact" (TIC) situation when he was hit with three rounds of AK47 in the chest. I knew after that that I was not responsible for his loss. I had constantly stressed on every mission that there was no target worth going below 1500 feet and certainly not below 1000 feet at any time intentionally. This young LT evidently had ignored my words and felt he was bulletproof and that he could win the war on his own. The problem is that two people on the ground with AK47s have you outgunned. They also are stationary while the FAC is flying while firing his weapon out the window. My personal guilt turned more to anger as I accepted his loss as one of being stupid rather than one of omission of my instruction.

The second loss was of a Captain from Laughlin AFB where he had also been an IP. Being from Laughlin AFB myself, I knew him pretty well. I also knew that he was planning on getting out of the Air Force upon return to the States and that he was going to go to law school. In fact, he had already been accepted for law school when he arrived in Vietnam.

Some of the FACs coming to Vietnam had been sent to language school at Monterey, California, to speak French. These FACs were to be utilized in Cambodia where French was more often spoken than English. This Captain was one of those who had completed

French Language School. After I completed checking him out in-country, I decided that I needed to send him to the FOL at Pleiku in MRII because I was lacking experience at this FOL. I was receiving feedback from individuals at Pleiku that some of the young Lieutenants were doing things that they were not supposed to be doing. I told him I needed him to go to Pleiku and rein in these young Lieutenants that were being cowboys. He literally begged me not to send him to MRII. He emphasized that he had been to language school and, therefore, he should fly in Cambodia. Cambodia was less of a risk for FACs as compared to MRII. I told him I truly needed him at Pleiku to provide discipline to the young Lieutenants at that FOL. He reluctantly agreed and went to Pleiku.

It was about a week before I left Vietnam when he and another FAC were reported killed in action. I truly took his loss personally since I had insisted that he go to Pleiku. I also knew he had a family and of his plan to attend law school upon returning to the States. His loss truly bothered me until two months later, when I was in graduate school at the University of Denver, and I found out that he and the other pilot had been doing a low pass and an aileron roll over a FB on his "finni" flight and had crashed. Instead of him providing the discipline to the Lieutenants, the young Lieutenants had won him over to the cowboy way. It was after that that my feelings once again turned to anger, and I no longer felt responsible for his loss.

I flew my last mission two days before I left Vietnam. It was on this particular mission, which was in MRIII, when I lost the rear engine for the third time. In MRIII, nevertheless, the topography is flat and maintaining 1000 feet did not represent a significant problem while returning to base. I returned to Tan Son Nhut and decided that was enough for me. This was the third engine that I had lost since being in-country. In addition, it was late May 1972, and the North's offensive had pretty much been halted.

In MRI, the South Vietnamese were in the process of pushing the North back into Laos and were regaining control of Quang Tri Province. In MRII, the North had been stopped from capturing a seaport and were retreating back across the border into Laos and Cambodia. In MRIII, the North had been stopped approximately

50 miles north of Saigon and were retreating. During this offensive in 1972, the South Vietnamese forces had successfully defended their country with U.S. assistance provided by U.S. Air Power and the few remaining troops/advisors along with U.S. financial support.

I left Vietnam on May 26, 1972, for the States. My tour had been cut short by about a month and a half due to my follow-on assignment. I had been selected for the Air Force Institute of Technology (AFIT) to attend the University of Denver in Denver, Colorado, to obtain a Master's Degree (MS) in Engineering. With the drawdown of the war, the Air Force did not need as many pilots to fly aircraft.

Since my undergraduate degree was Aerospace Engineering, I became a prime candidate to get out of the cockpit and get a MS degree in Engineering. I was scheduled to start classes in Denver in early June 1972. It was going to be tight to get back to Del Rio, Texas, and move my family to Denver and start classes in early June 1972. I was happy to be going home and quite satisfied with my efforts in Vietnam.

An interesting side note occurred when I got the AFIT assignment to the University of Denver. The assignment came down during the time period that I was working all day and flying all night. I was notified that I needed to take the SAT (Scholastic Aptitude Test) test for admission purposes. Here I was in Vietnam, in a very "hot" war, and going back to school definitely was not my primary focus. I also wondered where I could even take such a test in Vietnam.

I went over to Personnel and asked them if they could help me. They said that they thought that the Military Assistance Command, Vietnam (MACV), could help me out. I went over to their headquarters to see if they could provide me with the assistance that I needed. They indicated that they could not provide me with the SAT test, but they could give me the ATGSB test (Advanced Test for Graduate School for Business). I told Personnel that I could only get the ATGSB test and not the SAT. They said take the ATGSB. I scheduled the test as soon as I could get it.

It was scheduled during the time period that I was flying nights in support of An Loc. I truly do not remember much about the test, but at least I did complete it. My mind was elsewhere during that time period, and I was not sure that I was even going to make it home. I guess I passed, but I have no idea how I did on the test.

I departed Vietnam after completing 317 combat missions and 630 hours of combat time during the 334 days that I was there. I was looking forward to getting home, but I did not have a clue about the change in attitude of the American Society that greeted me upon my return. I soon found out how much the American culture and attitude had changed as far as the Vietnam War was concerned. It was also ironic that my follow-on assignment to graduate school actually set up the circumstance for my return to Vietnam in April 1975.

# CHAPTER 21

# The Beginning of The Long Return

UPON MY RETURN TO the States in late May 1972, I encountered my first impression of how significantly the American society environment had changed from when I had left. My first impression occurred upon my arrival at the San Antonio Airport. I was wearing my uniform when I exited the aircraft around 2200 hours (10:00 P.M.). My wife and oldest son, who had stayed in Del Rio, Texas, during my time in Vietnam, had driven over to San Antonio to pick me up. As I was entering the airport terminal area, I encountered the first anti-war sentiment toward those of us who had served in Vietnam. A young long-haired individual who was walking by me toward the gate area decided to spit on the floor in front of me. He did not spit directly on me, but the glaring look of disdain that he gave me as he spit on the floor in front of me told me that a significant change in the American general public's attitude had occurred since I had left. I did not say anything of the incident to my wife and son since I was happy to have survived Vietnam and be home. This incident did, however, leave me somewhat confused as to why this individual would have done such a thing.

My next impression of the significant amount of change that had occurred came as I watched the nightly national news reports on the major networks, especially Walter Cronkite, and their reporting on the war. It seemed that their emphasis was on reporting and showing all the protest that were occurring. It was totally obvious that the biased left-wing liberal news media was totally against the war. This was evidenced by the general anti-war sentiment that they were expressing. I soon discovered that this general anti-war sentiment was not only against the war but also against those of us who had done our duty to serve our country. My culture shock upon my physical return was steadily growing as I tried to process the change that I was experiencing.

My follow-on assignment from Vietnam was to attend the University of Denver (DU) in Denver, Colorado. The Air Force had assigned me there to get a MS degree in Engineering. I reported for classes in early June 1972. The University of Denver was and still is a quite liberal University.

My attendance in classes and interaction with the other students continued to reinforce my perception of the negative anti-war attitude of the general public toward the military and its members. This confirmed to me that indeed the general attitude of the United States had significantly changed. This negative attitude toward the military fostered a strong feeling of shame of my actions and accomplishments in Vietnam and even my status as an Air Force officer and pilot.

As a FAC in Vietnam, I had been awarded a number of medals. These medals included a Distinguish Flying Cross (DFC), 10 Air Medals, and the Vietnamese Cross of Gallantry with Palm. I received my medals by having them handed to me by another student in a class that he and I were taking. I did not know him, but he knew my name, and he happened to be in the Administration Office when my medals arrived at the University. He volunteered to give them to me since he would see me in class. This method of receiving my medals conveyed to me the idea that these medals did not truly mean much and truly did not matter. My efforts in Vietnam were truly not important. I also watched John Kerry throw his medals over the White House fence. And then I saw Jane Fonda sitting on an anti-aircraft gun in Hanoi where our POWs were. This was the last straw.

Based on the current environment that I was experiencing, I took my medals home and placed them in the bottom drawer of a steel filing cabinet that was located in a closet in our bedroom. The negative environment that I was experiencing led me to withdraw from any discussion of Vietnam with anyone including my family. I essentially shut down on the previous year of my life.

Going back to college in Engineering in 1972 caused me to have great concern about whether I would be able to compete with the younger generation of students. After all, I had graduated from Iowa State University in December 1964 and had immediately entered the Air Force and pilot training. It was now June 1972, and I had not truly even thought about college and the rigors of engineering. I also did not have any experience in the workforce utilizing engineering applications. I was concerned! This concern was erased in one of the first classes that I took. I experienced another incident that again confirmed how much the general attitude of the younger generation had changed.

It was in one of my first classes on the day before the mid-term exam was to be given when a longhaired hippie type individual tapped me on the shoulder. This individual would come to class barefooted, and he brought his dog with him. The dog (a German Shepherd) would lie at the back of the room and chew on a beer can. I always thought this was strange, but the instructor never said a word to, nor about, this individual. After he tapped me on the shoulder, I turned toward him, and he asked me, "Hey Dude, what does the textbook for this class look like?" I knew right then and there that I could compete with this younger generation.

This type of attitude was totally foreign to me, and I could not understand it. This attitude was so much different from the one I had experienced at Iowa State University (ISU) in the early 1960s. My undergraduate degree was in Aerospace Engineering, and none of my ISU classmates had an attitude like the one I had experienced. It did reinforce my feeling that I did not fit well into this different environment, but it did cause me to focus on school and not worry about others.

During the remainder of 1972, while at the University of Denver, I watched the anti-war, anti-military sentiment continue to grow. The United States Government was doing everything it could to withdraw from Vietnam. It also seemed that this anti-war sentiment was the main focus of the news media.

After that on January 27, 1973, our government signed the Vietnam Peace Accord with North Vietnam. A U.S. ceasefire went into effect on January 28, 1973, but we still had military advisors in South Vietnam assisting the South Vietnamese forces in defending their country. Even with this peace accord, the anti-war sentiment did not stop but continued by demanding that all of our troops come home. Finally, in February 1975, the anti-war supporters and demonstrators were able to apply enough pressure — along with enough anti-war politicians having been elected, that our government voted to cut off all financial aid to South Vietnam. This was when the Vietnam War was lost. The Politicians lost the war, NOT THE MILITARY!

During the Spring Offensive in 1972, the South Vietnamese forces had been able to repel the North with our financial aid and military support. Without our aid, it was only a matter of time before the North would attempt another takeover of the South. By cutting all financial aid, the U.S. indicated to North Vietnam that it no longer cared about South Vietnam and they were free to do as they pleased without interference from us. This action by our government in February 1975, which cutoff all financial aid was like a slap in the face to the military. It seemed to say that all of the actions and sacrifices of military lives did not matter.

This reinforced the perception to those of us who served in Vietnam that we should be ashamed of our actions, and we should not talk about it. It reinforced our actions of shutting down and internalizing the conflicting feelings that we had. It was hard for me to understand that we had lost over 58,000 U.S. soldiers during the conflict and now none of that seemed to matter. I had lost a number of friends, pilot training classmates and students whom I had trained as pilots — and for what?

I graduated from the University of Denver on December 7, 1973 with a MS Degree in Civil Engineering. The Air Force assigned me to Kelly AFB in San Antonio, Texas, to work as a Structural Engineer on the C-5A fleet of aircraft. I soon worked my way up to be the Lead Structural Engineer for the C-5A fleet for the Air Force. Kelly AFB was the System Program Manager for the C-5A fleet. My position as the Lead Structural Engineer is what resulted in my return to Vietnam in 1975.

One good event did occur while we were at the University of Denver and that was our second son was born. I did not want to leave Jan with two children if I did not make it back from Vietnam. He was born on August 11, 1973. This resulted in 7 years between our two sons due to Vietnam.

This assignment to the University of Denver was also more empirical evidence of God's plan for me. I was never asked what field of engineering I wanted to pursue. It was God who determined that I would pursue the MS in Civil Engineering. When I got to the University of Denver, I had to ask what field am I enrolled in. The Air Force had assigned me to the University of Denver for a year and one half. I only needed one year to get the MS in engineering since my BS was in engineering. With the extra six months, I completed the first year of the MBA program.

Upon graduation from the University of Denver, I was assigned to At Kelly AFB. At Kelly AFB I was able to complete the second year of the MBA program and become a Texas licensed Professional Engineer (PE). In addition, while at Kelly AFB, I was promoted to Major a year early due to my efforts in Vietnam. With this promotion, I was automatically selected to attend the Air Force Air Command and Staff College (ACSC) at Maxwell AFB in Montgomery, Alabama at the completion of my assignment at Kelly AFB. All of this was additional empirical evidence that God was guiding my life because it prepared me for life after retirement from the Air Force as well as the next important event in my life—"Operation Babylift." He was directing my assignments to do what he needed me to do as part of his plan for me and the second half of my career.

# CHAPTER 22

# Operation Babylift

SINCE OUR GOVERNMENT HAD cut off all financial aid to South Vietnam in February 1975, it was not long before the North launched another major offensive to take over the South. Without our aid and military advisors, it became obvious that the South would not be able to successfully defend itself as they had in 1972. The U.S. began a program of evacuating U.S. civilians and embassy personnel from Vietnam in April 1975.

It was on one of these evacuation missions from Vietnam when a C-5A crashed. This mission had been named "Operation Babylift." In addition to a number of civilian embassy employees who were being evacuated, the U.S. had agreed to evacuate a number of orphaned children. Many of these children had been fathered by U.S. soldiers and then abandoned by their mothers after our military forces withdrew. This C-5A crashed on April 4, 1975. Being the C-5A Lead Structural Engineer, I found myself headed back to Vietnam to investigate this accident.

The 4th of April 1975 started for me at 0500 hours (5:00 A.M.) with a phone call from the Kelly AFB Command Post. When anything significant occurred with a C-5A aircraft, as the Lead Structural Engineer for the C-5A fleet, I was on the list to be notified. The Command Post notified me of the C-5A accident in Vietnam. I arrived at my office at 0700 hours (7:00 A.M.) that day anticipating that I would be required to complete some activity as a result of this accident. I did not anticipate that I would be assigned to the Accident Board to investigate the accident.

At approximately 0900 hours (9:00 A.M.), the Command Post again called me to inform me that Major General Kelly (The Kelly AFB Air Logistic Center (ALC) Commander) had selected Major Russ Gregory (a Mechanical Engineer for the C-5A fleet) and me to be on the Accident Board. The Accident Board President was to be a 2-Star Major General from the Military Airlift Command (MAC). Having a 2-Star General selected to be the Board President indicated how important the investigation of this accident was to the Air Force. I do not know of any other Air Force accident investigation teams being headed up by a 2-Star General. There may have been, but I don't know of any.

My selection to this Accident Board turned out to be the second major potential crossroad in my Air Force career. I do not know if it was coincidence or not, but both of these crossroads had to do with personnel from the Military Airlift Command (MAC). The first major crossroad was in Vietnam with the Colonel Wing Operations Officer and the Airlift Wing at Tan Son Nhut. The second major crossroad was with this 2-Star General who also was assigned to MAC as the Deputy Commander for Maintenance for MAC.

The Commander of MAC, a 4-Star General, requested two engineers from Kelly AFB be provided to be part of the accident investigation team. This was due to the fact that Kelly AFB was the Program Manager for the C-5A fleet. The Command Post also informed me that a C-135 aircraft from MAC Headquarters at Scott AFB, Illinois, would arrive at 1200 hours (12:00 P.M.) that day to pick up Major Gregory and me to go to Vietnam. I immediately went home to pack a bag in order to get back to the base for the 1200-hour departure.

As luck would have it, my mother-in-law and her sister had arrived from Iowa to visit my wife and me. I rapidly packed my TDY bag and told my wife I was heading back to Vietnam, and that I did not know how long I would be gone or when I would be back. My wife was used to this, but my mother-in-law and her sister were not. My wife said, "Okay," but the looks on my mother-in-law and her sister were priceless. My wife told me later that after I left, they both wanted to know how often something like this happened. My wife simply replied that she was used to it and went about her business. My wife is a quite remarkable woman.

We departed Kelly AFB at 1200 hours (12:00 P.M.) on April 4, 1975, heading for Clark AFB in the Philippines. The Accident Board consisted of approximately 35 board members headed by the 2-Star Major General Board President. A full Colonel from MAC had been chosen as the Investigation Officer. We had to stop at Clark AFB in the Philippines, since when we arrived in Vietnam, we would be counted as part of the U.S. troop strength level allowed to be in-country by the Paris Peace Accord Agreement that was now in effect. Because of this Peace Accord, only seven of us from the Accident Board team would be allowed in-country at any one time in order to stay below the strength level allowed in-country by the Accord. This was the beginning of the difficulty in conducting this accident investigation.

The next day, the seven of us selected by the General to go, were flown to Tan Son Nhut AB in Vietnam on a C-130 aircraft to begin the investigation. We also were not allowed to remain in-country overnight, so we had to be flown back to the Philippines each night by the C-130. In addition to not being able to stay overnight, and the fact that only seven of us could be in-country, we were not allowed to wear any rank on our fatigue uniforms, and we could not carry any weapons.

After arrival at Tan Son Nhut, Air America helicopters of the Central Intelligence Agency (CIA), transported us to the crash site. We essentially arrived on site 24 hours after the accident had occurred. Parts of the aircraft were still burning due to the jet fuel that was still present. The CIA and South Vietnamese personnel were still in the process of recovering remains. The survivors had been rescued the day before after the accident.

The aircraft had broken into four sections on impact. These sections were the tail section, the troop compartment section, the wing section and the cockpit section. The cargo section of the fuselage no longer existed. The entire cargo section of the aircraft had been totally destroyed on impact. Only the troop compartment, which is located above the cargo area, 19 feet above the cargo deck floor, had survived the accident.

This troop compartment, which had 80 airline-style seats, was now sitting on the rice paddy in its final resting place. The cockpit section and this troop compartment section is where the survivors of this accident had been located. Some of the survivors included the pilot, copilot, and loadmaster. Without these survivors, we may never have been able to determine the cause of this accident, especially in the time period that was available and location of the accident.

We also did not know the actual number of souls that were on board. All we knew was that there were greater than 300 people including a number of babies. We were later informed of the actual number of babies and older children who were on board. We had 145 babies and 102 older children—some with physical handicaps.

The 145 babies had been placed in the troop compartment seats with as many as two or three babies to a seat. The 102 older children and at least 50 U.S. Embassy personnel along with other civilians had been seated on the cargo deck floor. These passengers had been strapped down with cargo ties since there were no seats in the cargo area. The entire cargo section where all of these people were located, no longer existed. All of these lives were lost when the fuselage cargo section was totally destroyed upon impact. What we did know is that 175 people did survive, but more than 150 people had been lost.

Upon our arrival in Vietnam, it was quite obvious that South Vietnam was going to be taken over by the North. The C-5A accident site was approximately 20 miles northeast of Tan Son Nhut AB. While at the crash site, we could hear the war going on and see smoke rising from many of the villages that were located relatively close to Saigon. In addition, the physical conditions at Tan Son Nhut had greatly deteriorated from when I departed in May 1972. The area that I had worked out of was almost non-existent due to the amount of deterioration that it had experienced. It definitely did not look good for the future of South Vietnam.

In addition to the ongoing war, the small number of Accident Board members allowed in-country at any one time, and the fact that we had to commute between the Philippines and Vietnam each day, we had other significant obstacles to contend with while conducting this investigation. The first was that the accident site itself could not be secured. Since we were not allowed to carry weapons, the South Vietnamese said they would address our security needs. The troops that we were assigned to provide security turned out to be 12-14 years old barefooted boys in tattered clothing with AK-47s. So much for security during the day, and there was no security for the site at night. This was also a good indication of the deteriorating quality of the South Vietnamese forces.

On the first day of our arrival at the crash site, the four sections of the aircraft were, for the most part, intact. The cockpit section and the troop compartment had not been disturbed. By the second day, however, all of the seats in the troop compartment had disappeared along with some of the cockpit instruments. By the third day, all of the headliners in the troop compartment and all of the oxygen masks that had deployed when the rapid decompression had occurred, were gone. By the fourth day most of the aircraft wiring and aluminum tubing had been taken, and exterior metal skin sections of the aircraft structure were now disappearing.

Without proper security, the local population was salvaging aircraft parts and metal to sell to junkyards or to keep for their own use. As many as 150-200 locals would be present each day salvaging parts as we tried to investigate the accident. The aircraft wreckage was disappearing right before our eyes, and the locals salvaging parts greatly outnumbered us. As the Lead Structural Engineer from Kelly AFB, I was responsible for disposal of the wreckage. Because of the locals and the war, this turned out not to be a problem.

Since we did not have a complete manifest of the individuals who had been present on the cargo deck, the State Department was placing a significant amount of pressure on us to identify the human remains that were present under the troop compartment of those individuals who had been seated on the cargo deck floor. As the structural engineer, I was given the responsibility by the 2-Star General to get to these remains in order to accomplish this tasking. The most obvious method was to try to flip the troop compartment over. To attempt this action, the CIA provided two helicopters with cables that could be attached to hooks on the bottom of their helicopters. Due to the weight of the troop compartment, however, we were not able to accomplish that without tearing the hook supports out of the bottom of the helicopters. Another course of action was required.

The next course of action was to cut out each section of the troop compartment floor, which was also the ceiling of the cargo area. With the removal of each 2'x4' section of flooring, we finally gained access to the remains that were underneath. One can imagine what we encountered once we gained access to this area.

We did, however, find a couple of nametags, some purses and billfolds, but not much else that would allow identification of remains. After two days of this activity, with minimal results, the State Department agreed that this effort would not produce the desired results. It took me approximately two months to get the stench out of my mind from this effort. Something like this does leave a lasting impact on one's mind in more ways than one.

After approximately a week of commuting between the Philippines and Vietnam, we finally got clearance for all 35 Accident Board members to be in-country and remain overnight. This helped immensely in our efforts to investigate the accident. With the complete Board present, we were able to conduct interviews of the survivors such as the pilot, co-pilot and loadmaster. The information that we gained was significant. We learned that at approximately 23,000 feet on climb-out during departure over the South China Sea, an explosive rapid decompression occurred. This explosive rapid decompression was caused by a catastrophic failure of a section of the aft ramp and pressure door, which departed the aircraft. When the ramp section and pressure door were blown out of the aircraft, the majority of the flight control cables were torn out along with the rupture of the hydraulic lines running to the rudder and elevator controls of the aircraft. The pilot was left with one aileron and the four engines, which had not been impacted, to control the aircraft. It was only with outstanding airmanship and the grace of God that there were any survivors from this accident. This information allowed us to focus our attention on the aft ramp area and the aft ramp locking system. We began searching for any parts from the remaining aft ramp section and any parts from the locking system that made it to the crash site.

The C-5A aft ramp locking system consists of seven locks on each side of the aft ramp. Each lock consists of a stirrup or clevis, which is attached to the ramp. These seven stirrups are U shaped and have a one-inch steel pin in the open end that is engaged by a bell crank hook which is located on the fuselage. These seven bell crank hooks are tied together by six tie rods that connect to a single hydraulic actuator. When the actuator is activated, it moves all seven bell cranks forward to engage the seven stirrups on the ramp to affect the locking activity. The six tie rods between the seven bell cranks have to be of a precise length to ensure all seven locks obtain an over center and a locked condition. The tie rod lengths are critical for this locking activity to be completed. We now knew we needed to find the 14 aft ramp locks which consisted of 14 bell cranks (hooks), 14 stirrups, 12 tie rods and two hydraulic actuators. We did, however, have a problem. Part of the aft ramp and pressure door were at the bottom of the South China Sea.

The Board President contacted the Navy, and being a 2-Star General, he did not have much trouble getting the Navy's assistance in this matter. We were able to provide the Navy with a relatively precise location as to where we thought the aft ramp section and pressure door should be located. The Navy went to work searching the floor of the South China Sea.

Meanwhile, the majority of the Accident Board members continued to search for the parts that we most desperately needed. We also were looking for the Crash Data Indicator Positioning Recorder (CDIPR), which is the black box for the C-5A aircraft.

The CDIPR is located on top of the tail section and will deploy if it experiences three G's or is submerged in water. This black box monitors all of the aircraft flight and engine parameters and has the last 30 minutes of cockpit voice recording. The CDIPR was not at the crash site, and we could only assume that one of the locals had found it and took it home to sell or use. It did look like a tape recorder. As I mentioned previously, the locals greatly outnumbered the Accident Board in retrieving parts from the crash site.

It was during our second week in Saigon, and the war was not going well for the South, when two South Vietnamese F-5E pilots out of Bien Hoa attempted to bomb the South Vietnamese Presidential Palace. Soon after that, the South Vietnamese President resigned. Things were truly now not looking good for South Vietnam.

During our investigation, those of us at the accident site would occasionally see NVA soldiers come out of the bordering tree lines and observe us. In addition, they would occasionally fire a round over our heads to let us know that they were there. I did not feel that they were going to attack us since, if they did, I think they thought that this might bring the United State Military back into the conflict. It is sad to say, however, if we had been attacked, I am now not so sure that this would have been the U.S. response based on the amount of anti-war and anti-military sentiment that was present in the United States.

I do, however, remember one particular day that I think may have been the closest that we came to being attacked. We were at the crash site, and it was during the time period that we could only have seven members present in-country. There were seven of us along with the normal 150-200 locals gathering parts. Around 1100 hours (11:00 A.M.) the locals started disappearing from the crash site as did our security guards. As we looked around, there were only us seven Board members left. Having been a FAC, I had been given the responsibility of handling our radio. We had a radio for contact with the CIA Air America pilots if we needed it.

With us seven being the only ones left at the crash site, it was not difficult for me to convince the 2-Star General that something was not right. He agreed, and I contacted the Air America helicopter pilots and requested an immediate extraction from the crash site. Within five to seven minutes, two helicopters were there and we departed the crash site uneventfully.

The next day when we returned to the crash site, the North Vietnamese had broken the rice paddy dikes and reflooded the rice paddy where we were working. We had closed the dikes to drain as much as water as possible from the rice paddy to assist in locating parts. With the reflooding of the rice paddy, the water hampered our search for parts. In many places the water was thigh deep. This caused us to have to use metal polls to tap the ground until we contacted a metal object. The locals were doing the same thing.

An interesting incident occurred one day while we were contending with the water problem. One of the local young boys, who was also searching for metal parts evidently, contacted an unexploded ordinance that went off. The blast blew off his foot. An adult man carried this boy over to me and indicated he needed our help. Since I had the radio, I called Air America and requested a medevac extraction for this boy. An Air America helicopter arrived and medevacked the boy back to Tan Son Nhut.

I do not know what happened to this boy, whether he survived or not, but I thought it was interesting that the locals brought the boy to the Americans for help. After the man handed me the boy, he turned around and walked off. He seemed to know that we would take care of the boy. The image of this boy still remains clearly in my mind 50 years later.

Since the locals had removed a significant number of parts, and we had not found the CDIPR and we still needed key parts from the aft ramp, we initiated a program as an attempt to recover these key items. We began searching junkyards and other salvage facilities where the locals may have taken the salvaged parts to sell. We also printed "pointee-talkie" sheets with pictures of the parts that we were looking for along with how much we would pay for them. We began going around to these junkyards showing the "pointee-talkie" sheets. This effort did provide a good result.

A French reporter contacted us and indicated that he could get the CDIPR for us. We had indicated that we would pay $200,000 piasters, which was about $200.00 in American money. The French reporter did provide us with the CDIPR but it was not in good condition. It was obvious that someone had tried to get the tape recorder to work, but did not succeed. The cockpit recording tape was there but not on the take-up reel. We were, however, able to send the tape and the black box to the Federal Bureau of Investigation (FBI) Laboratory in Washington, D.C., for analysis. They were able to retrieve most of the information that we had hoped for from the black box.

We investigated the C-5A accident as best we could under the deteriorating war conditions that were occurring. Without the U.S.'s financial aid that had been cutoff in February, the South Vietnamese government was collapsing around us. Congress also did not approve the 720 million dollars of emergency aid that President Ford had requested when this offensive began. The U.S. had turned its back on South Vietnam.

We were very lucky in that two days before we left Vietnam, the Navy found the aft ramp section and pressure door at the bottom of the South China Sea. Once we got these sections, we had enough of the pieces of the puzzle to determine the cause of this accident.

Since it was obvious that South Vietnam was going to fall to the North, the Accident Investigation Team packed up the recovered aircraft parts that we had at that point and left. We departed on the last two fixed-wing aircraft to leave Tan Son Nhut Air Base. These two aircraft were C-141s that the 2-Star General had ordered to transport the Team and the recovered parts out of Vietnam. The next day the North Vietnamese cratered the runway.

We left Vietnam on April 27, 1975, and South Vietnam fell to the North on April 30, 1975, three days after our departure. Needless to say, even though I was responsible for disposal of the wreckage, the locals essentially had taken care of that. What was left was left to the North Vietnamese. The only wreckage retained were the two C-141 aircraft loads that we left with from Vietnam.

The Accident Investigation Team left Vietnam and landed at Clark AB in the Philippines. We felt that we had enough information to start analyzing the parts that we had. The Navy's recovery of the aft ramp section and pressure door from the bottom of the South China Sea provided us with a tremendous amount of key information as to the cause of this accident. The section of the aft ramp that the Navy recovered had three of the aft ramp stirrup (clevises) still attached. Two of these stirrups were in perfect condition while the third had only minor damage. These were the first three locks on the right side of the aircraft's aft ramp. This pointed to the idea that at least two of the locks had unlocked, and the third was also in the process of unlocking when the catastrophic failure of the aft ramp occurred. We had found the other four right side aft ramp locks at the crash site. They were all significantly damaged.

The aft ramp locking system on the C-5A is designed so that the aft ramp and pressure door can withstand the loads from the internal aircraft pressurization forces with three of the locks unlocked as long as these three locks are not consecutive. If three every other locks are unlocked, the remaining four locks can maintain the pressurization loads experienced on the ramp generated by the aircraft pressurization system. In this case, it appeared that the first three locks on the right side of the aircraft had unlocked resulting in catastrophic failure of the aft ramp. The unlocking of these three locks must have occurred as the aircraft was climbing through 23,000 feet as the cabin pressurization forces were building.

When these three locks unlocked, the pressurization forces being held by these three locks were immediately transferred (dumped) to the remaining four locks. With this condition, the surface of the aft ramp was not able to withstand all of the forces from the four locks; and catastrophic overload failure of the aft ramp occurred between the third and fourth lock. The aft ramp split apart due to the pressurization forces, and the right side of the ramp was blown downward as it pivoted around the left side that remained locked.

Since the pressure door is connected to the aft ramp, the right side of the pressure door also rotated downwards. The pressure door has four large steel finger supports that rest on four large stainless steel rollers that are located on the fuselage at the top and end of the cargo area. This area of the fuselage above the rollers contains the flight control cables and hydraulic lines that go to the tail section of the aircraft.

As the aft ramp split and pivoted around the left side locks, it pulled the pressure door down with this pivot action which caused the steel pressure door fingers to be pulled down off the steel rollers from right to left as the ramp and pressure door pivoted down and away from the aircraft. The steel fingers of the pressure door are what caused the damage to the aft fuselage area where the flight control cables and hydraulic lines are located going to the tail section. This all occurred in a second or less when the three locks unlocked and dumped their loads onto the remaining four locks. This is why an explosive rapid decompression occurred in the aircraft. We now understood why the explosive decompression had occurred and the failure pattern of the aft ramp and pressure door. We also knew why the flight controls and hydraulic lines had been disabled. This failure pattern did, however, raise a number of other questions.

One question was, why did this occur on the third takeoff after leaving Travis AFB where it had started this mission? The aircraft had landed once at another location before proceeding to Vietnam. The aircraft had picked up a load of 105 Howitzers to take to Vietnam. This failure did not occur on the second takeoff, so why did it occur on the third takeoff? These were questions we needed answers to, and we needed to go to Travis AFB to get the answers.

We had ruled out sabotage in Vietnam with bomb-sniffing dogs. We needed to rule out the possibility that a bomb may have been placed in one of the passenger personal bags which had been loaded prior to takeoff from Tan Son Nhut. Many of these bags had been placed on the aft ramp for the trip.

After a week at Clark AFB in the Philippines, the Team was now on their way to Travis AFB, California. This is where the aircraft had been assigned and where it started this fateful mission. We needed a lot more information in order to answer our questions. We needed to review the aircraft's 781 Form, which is the aircraft's maintenance records, and to talk to the maintenance personnel. We were beginning to piece together what had caused the failure sequence, and we needed to know why the failure had not occurred on the earlier takeoff from Travis AFB or the takeoff from the intermediate stop.

Review of the aircraft's 781 Form revealed that this particular aircraft had been in what is called a "Cann-Bird" status prior to its initial takeoff from Travis AFB. What this means is that this aircraft was being utilized as a spare parts store for maintenance. When supply did not have a spare part available, the part was cannibalized from this aircraft to be used on another aircraft.

During the C-5A procurement process, the program was having both cost and weight-overrun problems. In an attempt to stay within the budget, cuts were made. Many of the cuts made were by not buying all of the spare parts needed to maintain the aircraft. The idea was to buy the spare parts at a later date. To off-set the lack of spare parts, MAC had instituted a program whereby one aircraft would be placed in a status of non-flying in order for parts to be cannibalized for a time period, after that replacing all of the cannibalized parts in time to fly the aircraft as required. MAC was required to fly its assigned aircraft at least once every 30 days. This was the aircraft used as the "Cann-Bird" for 20 days, and after that during the last 10 days, maintenance personnel would replace all of the cannibalized parts to get the bird ready to fly. The trip to Vietnam was this bird's first mission after coming out of "Cann-Bird" status.

Review of the 781 Form revealed that 26 cannibalizations had been completed on this bird prior to putting the parts back on the aircraft. Two of these cannibalizations involved the removal of the two tie rods located between locks 2 and 3 and 3 and 4 of the right-side aft ramp locking system. This was a very big red flag. We now knew we had to interview the maintenance personnel who had replaced these two tie rods.

It is important to understand how the C-5A aft ramp locking system was designed to work. As I have noted, the system is a gang locking system of seven locks with one hydraulic actuator. When the actuator is activated, it drives the bell cranks (hooks) forward to engage the stirrups to a position that is 3° to 7° over center as the locked position. All seven locks have to reach the 3° to 7° over center position to be considered locked.

To do this, the six tie rods between the seven locks have to be of a precise and specific length between the bell cranks to ensure that the proper over center position is achieved for each lock. To ensure that these tie rod lengths are of the proper length, the aft ramp locking system has to be properly rigged to ensure that all seven locks reach the desired over center and locked position. If any single locking component in this system is removed, all seven locks and six tie rods have to be re-rigged to ensure that the desired over center and locked position of 3° to 7° is obtained for all locks.

This re-rigging maintenance procedure is a long and tedious procedure to say the least. In fact, the re-rigging procedure for this maintenance activity requires 35 pages of instructions in the C-5A Aircraft Maintenance Manual—the T.O.–C5A–2-12 Manual. It also requires approximately eight hours to complete the re-rigging procedure for each side of the ramp.

Discussion with the Maintenance Personnel who had performed the replacement activity to these two tie rods revealed the following. One Maintenance Mechanic had started the tie rod replacement procedure approximately halfway through the maintenance work shift. A second Mechanic on the next shift finished the job and had assumed that all the re-rigging instructions had been completed on the first shift. The first mechanic, however, had merely replaced the tie rods but had not completed the re-rigging procedure to verify that the tie rods were of the proper length between the locks.

The way the system works is that if the locks are between the 3° to 7° over center position, when the aircraft pressurization forces begin to build on climb-out, the forces being developed in the tie rods actually tries to force the locks into a greater over center position. If the locks are not over center, in the 3° to 7° position, when the aircraft pressurization forces begin to build on climb-out, the forces being developed in the tie rods actually try to drive the locks to an unlocked position. This is why the re-rigging procedure is so critical when the ramp locking system is disturbed.

Some may ask, "Is a gang locking system the best way to design a system, or is it better to have individual actuators for each lock?" Of course, the answer is, "Yes, a single actuator for each lock is best." When the C-5A was being designed, however, it was having problems with weight during the design phase. By using a system that utilized only one actuator for seven locks, the weight of the six additional actuators on each side of the ramp was saved.

It should be noted, that due to the weight problem during the procurement process of the C-5A aircraft by the Air Force, Lockheed Aircraft Company was given a $300.00 incentive for every pound of weight it could save. The savings of six hydraulic actuators per side provided a large incentive to persuade Lockheed to use the gang locking system for the C-5A. Although not the best design for a locking system, it is an adequate design if properly maintained. It does, however, require a lot of maintenance activity to maintain such a system. This locking system was also the beginning of the conflict between Major Russ Gregory and me and the 2-Star General Accident Board President.

At Travis AFB we found out that the aft ramp had been locked when the bird was placed in "Cann- Bird" status. We also learned that since the aft ramp re-rigging procedure had not been completed, the aft ramp had not been opened until arrival in Vietnam where the 105 Howitzers were unloaded. At the intermediate stop after leaving Travis AFB, the 105 Howitzers had been loaded through the front visor loading ramp system. The aft ramp had not been utilized until Vietnam when it was opened to unload the Howitzers. This meant that the aft ramp had not been utilized since prior to the cannibalization activity of the two tie rods when the bird was in "Cann" status. This explained a lot as to why the failure had not occurred on the first two takeoffs from Travis AFB and the intermediate stop.

The interview with the Load Master after the accident indicated that he did have some problem getting a locked condition at Tan Son Nhut prior to takeoff. He stated that it took him three or four times to get a locked green light indication when he closed the aft ramp at Tan Son Nhut. This did cause him some concern, but he was able to finally get a green light locked indication for the ramp.

Each lock has a squat switch that is supposed to indicate when each bell crank has achieved an over center (3° to 7°) and locked position. These squat switches, however, are mounted on the fuselage by thin metal clips. These clips are easily damaged or bent.

While at Travis AFB, the 2-Star General Board President began to hold daily meetings at 0900 hours (9:00 A.M.) each day to discuss the progress of the investigation. It was during these daily morning meetings that Russ Gregory and I got in trouble with the 2-Star General.

Each day the 2-Star General would come into the room and go directly to the blackboard and with a piece of chalk, he would write in large letters, "Design Deficiency." The discussion as to the actual cause of this accident was beginning to manifest itself in two possible scenarios. The scenario that the General was supporting was "Design Deficiency," due to the gang-locking system, which only had the single actuator. He felt that each lock should have had its own actuator.

The second theory about the cause of the accident was developed by Major Gregory and me along with discussion with engineers from Lockheed. We contended that the cause of the accident was due to "Maintenance" as a result of not completing the re-rigging procedures on the right-side aft ramp locking system during the replacement of the two tie-rods—not "Design Deficiency." We felt that the design, although not the best, was adequate when considering the weight problem of the C-5A during procurement. Our theory as to the cause of the accident was based on how the locking system actually works.

Since the aft ramp was locked and closed when the aircraft was placed in "Cann-Bird" status, all seven locks were in the over center position of 3° to 7°. The ramp was properly rigged prior to the last closure operation. Since the aft ramp had not been operated during the time it was in "Cann" status, the locks remained in the over center and locked position. This is why the ramp did not fail on the takeoff from Travis AFB.

In addition, since the front visor loading system was utilized during the loading of the Howitzers at the intermediate stop, the aft ramp again was not opened, and, therefore, the failure did not occur after the second takeoff. The aft ramp system was, however, utilized at Tan Son Nhut when the Howitzers were off-loaded. The aft ramp was opened to conduct the off-loading activity. This was the first time that the aft ramp system had been utilized since prior to the aircraft being placed in "Cann-Bird" status.

As I have previously noted, when the locks are in the 3° to 7° over center position, the forces developed in the tie rod system as the cabin pressurization builds on climb-out after takeoff, actually try to force the locks into a greater over-center and locked position. This is why there was no problem on the first two takeoffs. When the locks or some of the locks, however, are not in the 3° to 7° over center position or actually not over center at all, the forces in the tie rod system during climb-out are such that these forces try to drive the locks toward an unlocked position.

Our theory was that after the ramp was opened in Vietnam and after that closed prior to takeoff, the closure occurred with a non-rigged system. When this happened, the actual position of locks 1, 2, and 3 was now in doubt as to being in the 3° to 7° over center and locked position. This is why the Load Master said he had to try three or four times to get a green light locked indication at Tan Son Nhut prior to takeoff.

What we theorized was that there was an improper length of the two tie rods between locks 2 and 3 and 3 and 4 after their replacement, and the system was not re-rigged. We theorized that on the closing of the ramp and the attempt to lock the ramp, lock 1 did not achieve an over center position at all. We felt that lock 2 only achieved a position of top dead center and could go either way as the forces built up on climb-out. Lock 3 may have achieved a position of slightly over center, but it was less than 3° over center.

After takeoff and as the cabin pressurization forces were building, the forces developed in the tie rods between locks 1, 2 and 3 were building in such a manner as to drive these locks toward the unlocked direction. Since lock 1 was already unlocked and lock 2 was top dead center, as the forces were building, they were sufficient to pushed lock 2 to an unlocked position and also pull lock 3 to an unlocked position. When this occurred, the loads from locks 1, 2, and 3 were dumped onto the remaining four locks. This resulted in exceeding the structural limit of the aft ramp between locks 3 and 4, which resulted in catastrophic structural failure of the aft ramp. This resulted in the previously noted failure pattern and the explosive rapid decompression as the aft ramp and pressure door departed the aircraft.

We supported our theory based on the evidence of the perfect condition of the first two stirrups found on locks 1 and 2 on the ramp section retrieved by the Navy from the South China Sea. This, along with the evidence of minor damage to lock number 3 stirrup, further supported our theory. The 2-Star General did not agree with our theory and was pushing "Design Deficiency" as the cause.

The 2-Star General was a large man at 6'4" or 6'5", and with 2 Stars, he was quite intimidating by his physical appearance. As such, most of the Accident Board members were in tune to supporting the General's position. As engineers, however, Russ and I could not go along with his position. Especially me, since I was a Texas licensed Professional Engineer (PE).

Each day after the General wrote "Design Deficiency" on the blackboard, either Russ or I would raise our hand and express the position that we could not support that cause of failure. What followed after our expression of non-support was a verbal tirade from the General until he would end the meeting. Russ and I would take turns each day on who would raise their hand after which the tirade would occur.

This went on for a week with each 0900-hour (9:00 A.M.) meeting going the same way. After that on Friday of that week, after his verbal tirade, the General said that if we could not support him, he no longer wanted Russ and me on his Accident Board. Russ and I said that if we were dismissed from the Accident Board, we would have to write a dissenting report to his accident report. He responded with, "Do whatever you think you have to do." Russ and I departed Travis AFB and returned to Kelly AFB. Since the C-5A wreckage parts were the responsibility of Kelly AFB, we essentially took our parts and went home.

We had now placed ourselves in a situation where two Majors were disagreeing with a 2-Star General. Usually, in cases like that, the two Majors do not fair very well. Our careers were on the line. We could only hope that we could get the support of our 2-Star General, the Air Logistics Center (ALC) Commander, at Kelly AFB. In addition, he would have to get the support of the 4-Star General who was the Commander of the Air Force Logistics Command (AFLC). We knew that the 2-Star General Board President would have the support of the 4-Star General Commander of the Military Airlift Command (MAC).

The background history of the Board President 2-Star General does provide some insight into why he was so determined to reject our theory. I felt he had been placed in a no-win position when he was selected to be the Board President for this accident. When this 2-Star General was a full Colonel, he was the Air Force's C-5A System Program Office (SPO) Director that was responsible for procuring the C-5A aircraft for the Air Force. The C-5A SPO office was part of the Air Force Systems Command (AFSC). As the C-5A SPO Director, he was definitely aware of the weight problem of the C-5A and knew about the weight reduction $300 per pound incentive provided to the Lockheed Aircraft Company. In fact, as the C-5A SPO Director, he probably had to approve this reduction in weight for the single hydraulic actuator locking system.

Once the aircraft fleet was procured, the C-5A fleet was assigned to an operational command. The C-5A fleet was assigned to the Military Airlift Command (MAC) for operations. In addition, once an aircraft system has been assigned to an operational command, the program management of that fleet is transferred from AFSC to the Air Force Logistics Command (AFLC).

The Air Logistics Center (ALC) at Kelly AFB was assigned as the Maintenance Depot and Program Manager for the C-5A fleet. This responsibility for the C-5A fleet after that rests with the assigned ALC until the aircraft system is retired for service. Being assigned to Kelly AFB in support of the C-5A fleet is how Major Russ Gregory and I became involved with this accident.

The problem for the 2-Star General resulted from the fact that at the time of the accident, he was the MAC Deputy Commander for Maintenance for the entire command. This meant that he was responsible for all maintenance activities performed by MAC personnel. So the problem for this 2-Star General was to decide if he was not doing his job during the C-5A procurement process or that he was not doing his job currently as the MAC Deputy Commander for Maintenance. I think he decided to support the idea that he was not doing his job as C-5A Director as opposed to not doing his job now. I also felt that as a 2-Star General, he was not used to having people telling him "No."

Russ and I returned to Kelly AFB and, of course, had to report to General Kelly. General Kelly was the 2-Star General in command of the San Antonio ALC. He was also a large individual physically at 6'4" or 6'5" himself. This also made him a quite intimidating individual by his physical appearance. In addition, he had been a WWII Prisoner of War (POW). Russ and I both had a lot of respect for General Kelly.

We briefed General Kelly on the accident investigation and on the two theories as to the accident cause. Fortunately for us, General Kelly felt that our theory was the proper cause of the accident. We also informed him we had been dismissed from the Board by the 2-Star General Board President for non-support. General Kelly said that he would support us submitting a dissenting report to the Accident Board's report. In addition, since we had all of the wreckage parts, General Kelly made a vacant hangar available to us and told us to prove our theory, which after that became known as the San Antonio Position.

I am not sure but I always had the feeling that General Kelly did not truly care for the 2-Star Board President. He never said anything negative about him but his body language when he spoke of him revealed a dislike of the man.

Russ and I began our efforts to prove our theory of the accident cause. We ran numerous laboratory metallurgical analyses of the failed wreckage parts to determine the type of failure mode for each of the failed surfaces. We also built a mock-up of the aft ramp locking system in order to test our theory of the forces developed in the tie rods as cabin pressurization forces built on climb-out after takeoff. This was necessary in order for us to prove our theory as to position of locks 1, 2, and 3 after ramp closure and why these locks were unlocked or became unlocked in order to start the failure sequence. Our efforts proved successful, and we were able to duplicate the failure sequence to within 300 feet of the actual altitude that it occurred. With General Kelly's approval, we began writing our dissenting report for submittal to the Air Force Safety Office at the Pentagon for their consideration. The Air Force Safety Office would have the final say as to the cause of the accident.

When the 4-Star General Commander of MAC was informed that San Antonio was indeed going to submit a dissenting report, things became quite uncomfortable for Russ and me. General Kelly assured us that he had our backs and not to worry. We now had a situation where two Majors had said "No" to a 2-Star General, and another 2-Star General was telling a 4-Star General that he agreed with the two Majors. The lines had definitely been drawn in the sand. Before we completed our report to be submitted to the Air Force

Safety Office, the 4-Star MAC Commander requested a meeting of all parties involved. This meeting was held at Wright Patterson Air Force Base in Ohio. The situation was now at the 4-Star General level.

This meeting consisted of the following individuals: the three 4-Star General Commanders from MAC, AFSC and AFLC, our 2-Star General and the 2-Star General Board President. In addition, Russ and I, the 2 Majors, were also present. The five General officers sat at a round table in a small meeting room. Russ and I sat in chairs behind General Kelly. There were 16 stars in that room along with two lowly Majors. This meeting was quite intense. Talk about intimidation, this was it.

The 2-Star Board President started the meeting off with his version of the accident cause and provided his justification. General Kelly after that expressed the San Antonio Position as the cause in a very calm manner without once looking at the Board President. He informed the 4-Star Generals that our mock-up of the aft ramp locking system had verified the failure pattern and that we had been able to duplicate the failure sequence. This, of course, got to the Accident Board President, who began to raise his voice in rebuttal. The one comment that I distinctly remember him saying was, "I don't care what those two Majors say; the cause was "Design Deficiency." Russ and I were not asked to say anything. General Kelly did all of the talking for us.

The results of this meeting ended with the MAC Commander supporting his Board President. The AFLC Commander expressed support for General Kelly and the San Antonio Position. The AFSC Commander did not support either side. I think he wanted to wait to see what the Air Force Safety Office had to say. If the Safety Office accepted the Board President's version, the cause would be focused on him and AFSC. If the San Antonio Position was accepted, the cause of the accident would be focused on MAC and not AFSC. He decided to wait and see. This resulted in what could be called a "Hung Jury." We all returned to our respective bases with no resolution as to an agreed cause of this accident. Russ and I finished our report, and General Kelly submitted it to the Air Force Safety Office at the Pentagon. All we could do now was wait to see which way the Air Force Safety Office would rule.

This turned out to be a long six-month wait for their final ruling, especially for me. Russ Gregory had been enlisted before going to AFIT to get his engineering degree and commissioning. He had reached retirement eligibility during this time period and retired from the Air Force. This left me alone awaiting the answer from Air Force Safety, which would determine if I had a career to finish. I was not eligible for retirement and would not be for a while.

The waiting was quite stressful, and I had to wonder if we had chosen the correct path. Majors normally do not do well when saying "No" to Generals. Generals are quite used to getting their way on things. With 4-Star Generals involved, I was not sure how it was going to turn out. Politics play a large roll in relationships among General officers. I hoped that the Air Force Safety Office would not play politics.

The thickness of the Accident Board's Report consisted of two volumes each three inches thick. The San Antonio dissenting report consisted of one volume that was three inches thick. This was a lot of information for the Air Force Safety Office to review and digest.

After approximately six months, General Kelly received a message from the Air Force Safety Office, which simply said, "We concur with the San Antonio Position as to the cause of the C-5A accident in Vietnam on 4 April 1975." General Kelly called me at my desk and read the message to me. The weight of the world was lifted off my shoulders. I knew that I now would be able to complete my Air Force career. To me, this was another example of evidence that God was guiding my path of life.

I did feel sorry for the 2-Star Board President. He was given one week to retire and lost a star as well. He retired as a 1-Star General, which is not bad. I did feel sorry for him because I think deep down he did agree with our version as the cause but thought he could sell his version in order to deflect the blame from MAC. In most cases, Generals do get their way. The Air Force Safety Office, however, does not play politics especially when loss of life is involved. Their goal is to make flight operations as safe as possible and to prevent similar accidents from happening in the future.

With the Air Force Safety Office's ruling, it was both good and bad for me. As I said, it was good in that I now had the opportunity to finish my Air Force career. It was bad in that now all of the lawsuits that had been filed against the Air Force as a result of this accident now rested solely on me as the Air Force's only authority representative. Russ had retired, and now I was the only one left to defend the Air Force. I think this is why God wanted me on this Board so that the actual cause of the accident would be determined.

For the next 10 years, each time one of these lawsuits came up, I had to go to Washington, D.C., to give a deposition on behalf of the Air Force. This occurred while I was at the Air Command and Staff College, after I left Kelly AFB, and numerous times while I was stationed at Cannon AFB flying F-111s. At least while I was at Cannon AFB, I could take an F-111 to Washington, D.C., and return. This kept my time away from my primary duty to a minimum; however, I had to keep a steel file cabinet full of my accident investigation efforts at home.

Being the Lead Structural Engineer at the time of the accident, I was also the engineer responsible for developing a sure-fire method of insuring that all of the locks did obtain the 3° to 7° over-center position during the locking activity. Essentially, we modified the C-5A ramp locking system by requiring large steel safety pins to be installed in each bell crank after locking. This "Murphy Proofed" the system. As I said, the gang-locking system was adequate but not the best. The C-5A ramp locking system is a quite complex system to properly maintain. The steel pins removed all doubt as to whether the system was properly rigged. The Air Force has never had another C-5A accident as a result of the modified aft ramp locking system.

Note: A Canadian film company, NextFilm Productions, Inc., made a documentary film of this C-5A accident for the "National Geographic TV Network's series "Air Disasters," Episode 5/53, called "Operation Babylift." This can be found on the Internet.

# CHAPTER 23

# The Second Half of My USAF Career

As I MENTIONED EARLIER, having been promoted to Major "below the zone" automatically qualified me to attend the Air Force Command and Staff College (ACSC) at Maxwell AFB in Alabama. This is equivalent to getting a master's degree in Air Force doctrine. Being able to go in residence is a significant honor. After completing my assignment at Kelly AFB, I was assigned to go to ACSC in 1977, which was a yearlong course.

It was at ACSC that again I realized that God was guiding my life as additional empirical evidence surfaced. This is when Ray Leach from my Undergraduate Pilot Training (UPT) class had a significant impact on my life. Ray and I had kept in touch periodically throughout the years. We even made contact in Vietnam. He had also chosen an instructor pilot (IP) slot out of UPT. Being number one in the class, he got his choice of assignment. He elected to stay at Reese AFB in T-38s. He chose an IP slot because, being a captain and now a newly rated pilot, he knew that he had to acquire a lot of flying time in a short period of time in order to be competitive with his peers for promotion to major. I wanted it because I did not want to fly as a copilot in the back seat of an F-4. After he completed his IP assignment at Reese, he received an F-4 fighter assignment to Vietnam. He was stationed at Da Nang when I was at Hue Phu Bai flying O-2As. It worked for me because when I left Laughlin AFB in January 1971, I had over 2,500 total flying hours with over 2,000 IP hours and zero copilot time. I am sure he did as well.

While I was at ACSC, Ray was flying F-111s at Cannon AFB in Clovis, New Mexico. Ray had been able to stay in fighters after leaving Vietnam. He did this even though the Air Force did not have enough cockpits available for all the pilots as a result of the Vietnam War. Ray, however, knew the fighter assignment representative for Air Force fighter assignments at Randolph AFB in San Antonio, Texas. They had served together, so Ray made a few calls for me and got me as assignment to F-111s at Cannon AFB. I had always wanted a fighter, and this time, it was God's plan for me to get one. Only God could have guided me in the Air Force to arrive at this point, and the empirical evidence supports this.

I received an F-111 assignment to Cannon AFB in 1978. When I arrived at Cannon AFB, I was told that I would be there for two years and after that I would be going to Lakenheath AB in England. Lakenheath AB also had F-111s stationed there, and this was the normal follow-on assignment for F-111 pilots leaving Cannon AFB. Nevertheless, this was not God's plan for me. He had a different plan as far as where I would be flying the F-111.

Cannon AFB was the home of the Twenty Seventh Tactical Fighter Wing (TFW). There were four squadrons of F-111D aircraft at Cannon. The 481st Tactical Fighter Squadron (TFS) was a training squadron for the 27th TFW. There were two combat squadrons stationed at Cannon: the 522nd TFS and the 524th TFS. The other squadron was the 523rd TFS and was a replacement training unit (RTU). Their mission was to provide replacement training for all aircrews going to Lakenheath AB in England. Lakenheath AB had F-111F aircraft stationed there. The F-111D was the first digital aircraft for the Air Force, and the F-111F was the follow-on version of the digital aircraft. Ray Leach was the 523rd TFS Squadron Commander when I received my F-111 assignment.

After completing my checkout in the 481st TFS, I was assigned to the 524th TFS. This is when God's plan for me at Cannon AFB began to be revealed. Being a major, I was assigned as an assistant operations officer for the squadron. I was quite surprised as this was a high position for someone who had just become an F-111 pilot. I do not know, but I suspect that Ray Leach might have had a say as to where I was assigned.

As I was approaching the two-year point at Cannon AFB, I was expecting to receive an assignment to Lakenheath as I had been told would happen. God, however, had a different plan for me when he provided a new opportunity at Cannon AFB. The Wing Commander asked me if I would like to stay at Cannon AFB as the Director of Wing Training for the Twenty-Seventh TFW. I guess they felt that because of my UPT instructor experience and the fact that I had set up an O-2A in-country training program in Vietnam, I would be a good fit for them as the Director of Wing Training. This position also provided me with another opportunity, in that I was upgraded to an Instructor Pilot (IP) in the F-111. I know that God had to be guiding me because he had provided all the previous flying training experience in my previous assignments as he guided my life. This Wing position also froze me at Cannon AFB for another two years. In addition, I had been promoted to Lt. Colonel while I was in the 524th TFS.

After two years as Director of Wing Training, God provided me with another opportunity at Cannon AFB, which again froze me for an additional two years. The Wing Commander selected me to become the Operations Officer of the 523rd TFS. This was the RTU squadron for the training of aircrews going to Lakenheath AB in England. Since I was already an IP, I had no problem fitting into the 523rd TFS. It is kind of ironic, but I had now gotten back to the level of responsibility that I had as a captain when I was the Operations Officer of the 21st TASS in Vietnam. This is more empirical evidence of God's plan for me.

When the two years as the 523rd TFS Operations Officer was up, God provided me with another tremendous opportunity at Cannon AFB. When the 523rd TFS Squadron Commander position became available, I was selected to become its Squadron Commander. This was also ironic because this was the same squadron that Ray Leach had been the commander of when I arrived at Cannon AFB. Being selected for this position could have only been due to God's hand in making this happen. There were many factors involved that indicated that I would not be selected to become a squadron commander.

This first major factor was that all squadron commanders had to live on base. When I was first assigned to Cannon AFB, base housing was not available, so we had purchased a house in town. To live off base as a squadron commander required the approval of the Twelfth Air Force Commander, a three-star general stationed at Bergstrom AFB in Austin, Texas. This is where I know that God had to have intervened on my behalf because the three-star general approved my living off base. This was against normal Tactical Air Command (TAC) policy for their Squadron Commanders.

The second major factor for me to be approved as a Squadron Commander involved my wife. Jan had always worked during most of my assignments. She had her teaching degree in English and had taught English at Del Rio High School during our assignment at Laughlin AFB. Her teaching job in Del Rio was one of the reasons she stayed in Del Rio while I was in Vietnam. We had purchased a house in Del Rio again because of the lack of base housing at Laughlin AFB when we arrived.

While we were in Denver and I was going to school, she did not teach since we were only to be there for a year and a half. In addition, God blessed us with our second son while we were in Denver.

While we were stationed at Kelly AFB in San Antonio, Texas, she actually taught at two different high schools. Even when we were at ACSC College at Maxwell AFB in Montgomery, Alabama, she taught that year at a private school. When we arrived at Cannon AFB and thinking that we would only be there for two years, she decided to get her master's degree in guidance and counseling at Eastern New Mexico University. When it became apparent that we were not going to Lakenheath on the time schedule that we had been told, she got an opportunity to be the Director of Guidance and Counseling at the local junior college in Clovis, New Mexico. This was the job she had when I was selected to be the 523rd TFS commander. To me, this is empirical evidence that God was also guiding Jan's life, as he was mine, as he was providing opportunities for her.

In the Tactical Air Command (TAC), Squadron Commanders' wives are considered to be a very important aspect in the life and functioning of the squadron. They essentially have to be "Den Mothers" to the wives of all members of the squadron. Normally, wives of Squadron Commanders do not work because it possibly would interfere with the duties that the Squadron Commander's wife is expected to perform.

Naturally, Jan did not like the idea that she would have to give up her job. She felt that she was not the one in the Air Force; only I was. I truly could not blame her for feeling that way. She had worked quite hard to get to where she was, and she had an excellent job. Being in the TAC and flying F-111s, we deployed a lot, and her job helped her while I was gone. I also was not going to ask her to give up her job. Although, I knew she would have if I had asked her. That is the kind of wife she is.

This again is when I know that God intervened on our behalf. To become a TAC Squadron Commander, all candidates had to be interviewed by the three-star Twelfth Air Force Commander. The subject of my wife's working came up in my interview. The Wing Commander had briefed the three-star general that I lived off base and that my wife was working. The general asked me if she would be able to complete her duties as the Squadron Commander's wife. I knew that Jan did not want to quit her job, and I also knew she would perform her duties as the Squadron Commander's wife in an excellent manner.

I confirmed to him that my wife could do both: work and perform her Squadron Commander's wife duties. The three-star paused a moment and after that agreed to allow her to continue working. I think I was the first squadron commander to live off base and have a wife that worked. Only God could have made this possible. To me, this is more empirical evidence that God does exist.

As I reflect back over this time period, I feel that God was working to keep me at Cannon AFB beyond the two years that I was initially told for a specific reason. We actually stayed at Cannon AFB for seven years. The reason I feel that we stayed at Cannon AFB so long goes back to the C-5A accident and report. I feel that it was his plan for me to be on that accident board, and that was the reason he provided me with the opportunity to get my master's degree in engineering and after that be assigned to Kelly AFB. He also guided me to obtain my professional engineer (PE) license while at Kelly AFB and become the C-5A lead structural engineer. It was his plan that I was assigned to that accident board, and after that he guided me to stand up to the two-star general accident board president. This resulted in the correct cause of that accident being accepted by the Air Force.

Now I feel that God wanted me at Cannon AFB for the seven years so that I would be available to provide the depositions for the lawsuits against the Air Force as a result of that C-5A accident. If I had been in England at Lakenheath AB, there would have been significant delays on getting me back to Washington, DC, each time a deposition was required.

During the seven years that we were at Cannon AFB, I was called to Washington, DC, a number of times. When I was needed, the Wing Commander would get a call from the Pentagon, and then he would call me and tell me to take a F-111 and go to Washington to give my deposition. This made it quite handy and allowed me to be able to be responsive to the lawyers' requests. This saved time and did not detract much from my assigned duties at Cannon AFB. It would have been a lot more difficult if I would have to come from Lakenheath each time a deposition was needed.

During my time at Cannon AFB, I also know that God was watching over me and was guiding my life. We did not have any aircraft accidents in any of the squadrons that I was in during my time as Assistant Operations Officer, Operations Officer, and as Squadron Commander. There were aircraft accidents in the Wing, but none in any of the Squadrons that I was in.

It was during the time that I was assigned as the Director of Wing Training that a reorganization of the wing occurred. The wing decreased from four squadrons to three. This was due to the loss of a few aircraft. The 481st TFS remained as the training squadron for the Wing. The 522nd TFS remained as the primary combat squadron for the Wing. The 524th TFS, which was the second combat squadron, was disbanded as part of the reorganization. The

523rd TFS remained as the replacement training unit (RTU) for Lakenheath, but also picked up the responsibility as the second combat squadron. This meant that the 523rd TFS had instructor pilots and instructor weapons system officers (IWSOs) for the RTU responsibilities and also had combat crews assigned to fulfill the combat role that the 523rd now had for the Wing.

The reorganization was also completed to better utilize the aircraft and help the wing be able to maintain the aircraft more efficiently. This allowed the wing to be able to obtain and then sustain the level of combat mission readiness required for tactical fighter wings in the Air Force. The F-111D fleet of aircraft had always been hindered by a lack of spare parts, and being the Air Force's first digital aircraft, we could never reach the combat mission readiness level required by the Air Force. It was not until President Reagan increased the military budget that sufficient parts were procured for the F-111D that we were able to meet and sustain mission-readiness standards. During the time period that the F-111D aircraft had been assigned to Cannon AFB, the wing was never at a high enough mission readiness level to receive an Operations Readiness Inspection (ORI) from headquarters of the Tactical Air Command (TAC).

The Operations Readiness Inspection (ORI) exercise is when a wing is tested by the Tactical Air Command (TAC) to determine if it can perform its wartime mission responsibilities. The results of these ORI exercises determine if tactical wing commanders make general and if squadron commanders make full colonel. I was the 523rd TFS Commander when Cannon AFB received its first ORI exercise.

These inspections usually start at some very unusual time of the day or week. It is normal to think that an enemy is most likely not going to start a war on Monday at 8:00 a.m. Our ORI started on Sunday at 10:00 a.m. when no flying was scheduled. A team from the Tactical Air Command (TAC) landed unannounced at Cannon AFB and started our ORI inspection. Once started, the two combat squadrons had one hour to get everyone back to the base. In our case, it was the 522nd TFS and my squadron, the 523rd TFS, which were being tested. The 481st training squadron was not involved.

With the ORI starting on Sunday morning, and with no cell phones, pagers, or other modern-day communication devices available, we had to depend on our Squadron alert rosters. This roster was a paper list of all member of the Squadron listed by Flights. Each Flight Commander was responsible for getting his aircrews back to the base within the one-hour period required. It was my responsibility to contact each Flight Commander to get the alert roster started as soon as I was notified by the command post that the ORI had started. If this one-hour time period is not met, the Wing fails the ORI before it even gets fully started.

You can only imagine what it was like getting all of our aircrews back to the base with only landline phones and face-to-face contact. All four of my Flight Commanders and I were going around to the churches in Clovis looking for our Air Force personnel to get them back to the base before the hour was up. My living off base turned out to be an advantage for me. Both squadrons were able to get everyone back with God's help.

The next critical point was that both squadrons had to pack up all of their war reserve materials (WRM) that would be required if we were actually going to war. In fact, in a way, we actually were. We had twenty-four hours to complete this task for both of the flying squadrons. The maintenance squadrons were required to pack up all of our war reserve spare parts kits (WRSK) that would be required, as well as get all twenty-four aircraft in each squadron ready for flight. At the end of the twenty-four-hour period, both squadrons had to deploy all twenty-four of their aircraft to prove that we could actually do it and that all aircraft were actually mission ready. Both the 522$^{nd}$ and the 523$^{rd}$ squadrons completed this requirement.

The next phase of the ORI started twenty-four hours later after we had proven that we could deploy all of our aircraft. Each of the squadrons had to be able to fly forty-five sorties a day on their twenty-four aircraft for three days and after that continue operations by flying thirty sorties a day thereafter. Our normal flying schedule consisted of flying fifteen sorties a day for each squadron. The forty-five/thirty-sortie rate for each squadron was our wartime tasking if we ever did have to go to war.

To maintain this sortie rate for the F-111D aircraft was very taxing for both maintenance and the flight crews. It meant that most crews had to fly twice a day, and the aircraft had to turn two or three times in a twenty-four-hour period in order to achieve this sortie rate. This included any aircraft repairs needed after each flight as well as the loading of the weapons on each aircraft for each mission. For the F-111D aircraft, this was a significant challenge since our mean time between failures (MTBF) was only 6.8 hours. This sortie rate was a formative tasking for the F-111D.

I know that God had to be watching over us and was guiding our efforts. I did not have to worry much about my instructor pilots and instructor WSOs, but I did have to worry about my young and inexperienced combat crews. All of my combat crews were young lieutenants on their first assignments after undergraduate pilot or navigator training. I had approximately twelve of these young lieutenant crews.

It is interesting to note how fast experienced combat pilots were no longer present in the field. The US pulled all of its combat forces out of Vietnam in 1973. At that time, a significant number of fighter pilots had combat experience. Ten years later in 1983, in the 523rd TFS, we had only 4 percent combat-experienced pilots present. My Operations Officer and I represented that 4 percent. We both had been O-2A forward air controllers (FACs) in Vietnam. There were no other combat-experienced pilots available.

With the ORI being a test of our wartime capabilities, and with only myself and my operations officer with combat experience, the remainder of my squadron did not have any idea how taxing this exercise would be. This is where my experience as an operations officer in Vietnam during the Spring Offensive in 1972 paid off. This is another example of empirical evidence of God's plan for me. I knew exactly what it was going to take to ensure the 523rd TFS did well on the ORI.

I flew the first mission of our assigned missions because I knew I would have to work very closely with our maintenance personnel to match aircrews to aircraft. I know that many of the aircraft would not be completely mission ready for each mission due to the failure rate of the F-111 in order to fly forty-five sorties a day.

This meant that after the first day, I would no longer have the required crew rest to fly additional missions. I was going to ride with the maintenance flight line supervisor to decide which crews should fly which aircraft since many of the systems would not be available for use by the crew. There were always going to be aircraft discrepancies after each flight.

My job, therefore, was to match the aircrews that I knew had the capability of flying each mission with discrepancies that could not be cleared between the missions. To do this, I was on the flight line for 6:00 a.m. until 2:00 a.m. the next morning. I was able to get a couple of hours of sleep and after that be back on the flight line at 6:00 a.m. to start the next day of sorties. It was similar to my schedule in Vietnam during the 1972 Spring Offensive. I had to depend on my Operations Officer to run the Squadron while I was on the flight line. I was essentially matching aircrew capabilities to what systems were or were not available on each aircraft in order to maximize our best chance of completing a successful mission. I know that God had to be guiding me because on one specific mission, I had to make a very critical decision as to whether to let one of my young inexperienced lieutenant crews fly a mission in an aircraft which was in need of serious repairs.

This particular mission was scheduled to be a live drop of twelve Mark 82 bombs (five-hundred-pound bombs) on a target at a small range in Colorado that I knew this aircrew had never been to. I also knew that a TAC flight examiner aircrew would be chasing the aircraft since this would be a live drop of bombs. The aircraft that had been scheduled for this mission needed repairs to the inertial navigation system (INS), the weapons radar, and the weapons release computer. All these systems are needed for the WSO to assist the pilot to complete the mission. Unfortunately, all these systems were inoperative, so the WSO would be along for the ride. We did not have time to change aircraft because the bombs were already loaded on this aircraft. In order to make the time over target, this aircraft had to go. The crew that was scheduled for this mission was one of my young inexperienced lieutenant crews.

I was very familiar with both of these crew members. I had flown with each of them during their checkout to become combat ready. Without any of the technology in the right seat working, this F-111D was essentially a two-engine F-100 aircraft. The crew would have to navigate to the target area, and the pilot would have to drop the bombs manually without any technical support from the WSO.

I made the decision to send this young crew in this aircraft because we did not have enough time to load another aircraft, and I did know this crew. This decision was fraught with all kinds of possible disasters that could go wrong. Even the maintenance supervisor knew what this decision meant if things went south.

I met my young crew at the aircraft as they arrived to begin their preflight of the aircraft. I started our conversation with, "I've got some good news and some bad news." I told my young crew that the bad news was that they would not have an INS, a radar, or a weapons delivery computer for this mission. I told the pilot that he would have to make a manual visual delivery of the bombs. I told them that the good news was that I knew that they could do it. They both looked at me with a noticeable disbelief on their faces. I then said to them, "I have flown with both of you, and I know you are capable of completing this mission." I then said, "You guys get into the aircraft, and I will do your preflight."

As the crew taxied out for takeoff, the maintenance supervisor and I looked at each other, both thinking without saying it, What have we done? I had the same feeling that I had when I soloed my Iranian student Ammand in the T-37. I was again betting my wings.

The next two and one-half hours were quite tense for us as we waited to hear the results of this mission. We could listen to the command post frequency in our truck to hear the results of the weapons delivery when the TAC flight examiner crew reported back the results of live-drop missions. While we were waiting, I had the same type of feeling I had when I lost the rear engine at night while flying the O-2A in MRII above clouds and mountains as I was trying to make it to the coast in Vietnam. It also was similar to the feeling I had when we were waiting for the Air Force Safety Office to decide which cause of the C-5A "Operation Babylift" accident that they were going to accept.

When we heard the radio call from the TAC flight examiner crew and they scored the mission as "bull" (bull's-eye), I also felt the same sort of relief as I had on all the previous major decisions that I had made that turned out good. Only God's hand could have been guiding all these events.

When the crew returned to Cannon AFB and got out of the aircraft, I had never seen any other crew have larger smiles on their faces and walked as tall as they both did as they returned to the Squadron. They knew they had done something very special. I was quite thankful to God for all of his help on this mission.

At the completion of the ORI, which lasted for a week, we all waited to hear the results. With the ORI being the Wing's report card and this being the first ORI for the F-111D aircraft, it was extremely important that we had to do well. A lot of careers were on the line.

The overall ORI results for the 27th TFW were "excellent." The results for the 522nd TFS were also "excellent." The results for the 523rd TFS, however, were "outstanding." I was extremely proud, and I knew that God had guided me to this point. As a result of the ORI, the Wing Commander made general, and both the 522nd Squadron Commander and I were promoted to full Colonel.

At the end of my two years as a Squadron Commander and having been promoted to full Colonel, I also was selected to go to the Air War College (AWC) in residence. This, again, was an opportunity that God provided as not all colonels get selected to go to AWC in residence. The Air War College is like getting a PhD in Air Force doctrine. The Air War College is also located at Maxwell AFB in Alabama, where ACSC is located. It was 1985, and we were now on our way back to Maxwell AFB for the second time. The Air War College (AWC), like the Air Command and Staff College (ACSC), is also a yearlong course of study. Jan decided not to teach this time while we were at Maxwell AFB. She decided she would take the year off.

The Air War College is a stepping-stone to making general. Nevertheless, one usually had to go to the Pentagon after graduation if one wants to try for the "brass ring" (make general). Assignments to the Pentagon, unfortunately, are very political in nature. One needs a sponsor to push one's career ahead at the Pentagon. I never was much for politics, and I did not have a sponsor that I knew of. In addition, the cost of living for military members in the Washington, DC, area is quite significant. Based on the above, I truly did not want to go to the air staff at the Pentagon. I requested to go back to Kelly AFB in San Antonio, Texas, instead of going to Washington, DC. God must have agreed with my decision because my follow-on assignment from AWC was back to Kelly AFB.

It should also be noted that the last request for my deposition for the C-5A accident lawsuits occurred while I was at AWC. So for the time period from ACSC, my seven years at Cannon AFB, and my time at the AWC, I had been called to go to Washington, DC, at least ten to twelve times. I lost track of the total number of times that I was requested to go. The time period, however, was almost ten years from the first trip to the last trip. I do feel that it was God's plan to keep me in the States as part of his overall plan for this accident.

Since I had turned down my chance to go to the Pentagon and compete for a promotion to general, we wanted to go back to San Antonio and retire from the Air Force. We had been stationed there two times, and San Antonio felt like home to us and was where we wanted to live after the Air Force. This must have also been God's plan for us because I did receive an assignment to Kelly AFB in San Antonio, Texas.

With our assignment in hand and before graduation, Jan went to San Antonio to buy our retirement home. I could not go with her because of school, so I told her to get the house that she wanted. She did exactly that, and we became owners of a house that I had never seen. She did a perfect job, and we still live in that house today.

I know that God was guiding her because the realtor that she found was the father of the wife of one of my flight commanders while we were at Cannon AFB. He was super helping her, and they found the perfect house for us. The house was only nine months old and still under warranty. The original owners had been transferred to Houston for job reasons. This to me was more empirical evidence that God was guiding both of our lives.

When I arrived at Kelly AFB in June 1985, I did not know exactly what I would be doing. Upon arrival, I reported to the

Air Logistics Center's (ALC) Vice Commander, who was a one-star general. He informed me that I would be going back to the Materiel Management (MM) Directorate where I would be the System Program Manager for all of the Air Force's small aircraft systems being managed by Kelly AFB. To me, this was great and more evidence that God was behind this assignment. I became the division chief of the small aircraft division (MMS) of the Materiel Management (MM) Directorate for the Air Logistics Center (ALC).

I became the System Program Manager for the Air Force's fleet of T-37s, AT-37s, T-38s, F-5s, F-102s, F-106s, O-2As, and OV-10 aircraft systems. It was a great fit for me since I had flown essentially five of the weapon systems that I was now responsible for. The F-5 is essentially the fighter version of the T-38. The Air Force added a gun and rocket pods to the T-38, and it became a fighter. The AT-37 is an upgraded T-37 with T-38 engines without the afterburner. In addition, a minigun was added as well as wing pylons to carry small bomb loads. It was now also a fighter, and both were used in Vietnam. I did get a ride in an AT-37 in Vietnam when one of my fellow instructor pilots from Laughlin AFB received an AT-37 to Vietnam as his assignment. He was assigned to Bien Hoa, and we utilized the AT-37 in support of Cambodia.

I performed system program manager duties for approximately a year and a half when God provided me with an opportunity for advancement as far as job opportunities at Kelly AFB. I was selected to become the number two colonel in the Materiel Management Directorate. I was chosen to become the Resource Management Division Chief (MMM). This position put me in charge of all of the materiel management budget authority for Kelly AFB. I was responsible for approximately $5 billion dollars of the Air Force's budget authority. This was a lot of money back then as it is even now. My signature was the approval authority for the expenditure of funds based on purchase requests (PRs) developed by item managers at Kelly AFB. I had approximately five hundred item managers whose jobs were to procure spare parts for all the weapon systems being managed at Kelly AFB. In addition, I also had responsibility for the budget of the "black" programs that Kelly AFB was assisting in procurement for the Air Force. Only the two-star ALC Commander and myself knew how much budget authority I had for the "black" programs. At that time, the Air Force was procuring the F-117 and the B-2 aircraft weapon systems. This job elevated me to the level from which another opportunity to possibly make general was afforded me as a follow-on assignment after Kelly AFB.

I was being considered to go to McClellan AFB in Sacramento, California, to become the Maintenance Directorate Commander for the F-111 aircraft weapon system. McClellan AFB was the Air Logistics Center (ALC) and depot repair facility for all of the Air Force's F-111 aircraft. This position was a one-star general position, and it was considered to be a great opportunity, if I accepted it. I also knew that if I turned it down, I would have to retire from the Air Force.

Since I had flown the F-111 for seven years, I knew how much of a nightmare it was maintaining the F-111. It was a great aircraft to fly, but a real challenge to maintain. In fact, after I had been promoted to Colonel while at Cannon AFB, I was required to give up the 523$^{rd}$ TFS Commander's position. My class date for the Air War College, however, was six months away. Cannon AFB needed to find a position for me while I waited for my departure to the Air War College. It so happened that the Deputy Commander for Maintenance for the Twenty-Seventh TFW had been fired. He was a full colonel and was retired immediately. This left an opening at the Wing. Since I had been promoted to colonel, the Wing Commander designated me to become the Wing's Deputy Commander for Maintenance. During those six months, I truly learned firsthand how hard it was to maintain the F-111 aircraft. This experience was an important factor in my decision.

A second factor in my decision involved the cost of living in California. The cost of living in California was quite high compared to what Jan and I had always been familiar with. To me it was similar to the Washington, DC area and was one of the reasons I did not want to go to the Pentagon. In addition, I felt that God was leading me in a new direction.

I turned the job down and retired from the Air Force on June 30, 1989. I was quite satisfied with my Air Force career, and I truly did not have any aspirations to become a general officer. I knew that to be a general, you had to be political in all of your dealings. Being an engineer, I knew that I was not a politician by nature; and besides, Jan and I were in San Antonio, Texas, and this was home.

# CHAPTER 24

# Postmilitary Years

LIFE AFTER THE AIR Force was going to be a new experience for both of us. I was approaching this new phase of my life with some anxiety about what I would do. God, however, had prepared me with all the pieces of the puzzle for my life's path that he had planned for me while I was in college and after that the military. These pieces of the puzzle were now becoming clear to me.

He had led me to choose Iowa State University and study aerospace engineering and to choose advanced Air Force ROTC. In the Air Force, he provided me with the opportunity to get a master's degree in civil engineering and an MBA degree. He also provided me with the opportunity to work as an engineer and even have the opportunity to become a Texas-licensed professional engineer (PE). Because of this, I was well prepared for my postmilitary career. It became obvious that God's plan for us was to retire in San Antonio after my second assignment to Kelly AFB.

After I retired from the Air Force, I began looking for an engineering position. I had a number of interviews with local San Antonio engineering firms. While I was looking for a job, God led me to the engineering firm that he had planned for me as part of my life after the Air Force. This firm was Chenn-Northern Inc., a geotechnical engineering firm with headquarters located in Denver, Colorado. They had a number of offices throughout the northwest portion of the United States. Their office here in San Antonio was the office farthest south and east of their headquarters in Denver.

The San Antonio Chenn-Northern office had experienced a mutiny by the office manager and a number of employees. The office manager and some of the employees had decided to start their own engineering consulting firm and had quit Chenn- Northern quite unexpectedly. When they left, they took a number of large clients like Exxon, Texaco, and Chevron with them. This had a significant impact on the San Antonio Chenn-Northern office, and they were looking for a new office manager. A Master's Degree in Civil Engineering and a Texas-licensed Professional Engineer (PE) was required and it was desirable to have an MBA degree. I was a perfect fit for them. I was hired in August of 1989 as their new Office Manager. This is all more empirical evidence of how God does have a plan for all of us.

After I was hired and began to see how much damage had occurred to the business after the previous office manager and the employees left, I knew it was going to be a real challenge to recover the Chenn-Northern office. This is when God guided me as to what exactly I needed to do.

This was during the time period when national environmental concerns were coming to the forefront and the main focus of the Environmental Protection Agency (EPA). There was significant focus on gasoline underground storage tanks (USTs) at all gasoline stations. The State of Texas had issued regulatory guidance requiring that all gasoline USTs had to be tested for leaks and even removed if over twenty-five years old. If leaks were found in the USTs that were less than twenty-five years old, after that the site had to be evaluated for hydrocarbons. If a subsurface hydrocarbon impact was identified at the site and if those hydrocarbons were detected in the soil and/or groundwater above the established regulatory levels, the owners were required to remediate the site.

The saving grace for me was that the Chenn-Northern office had a drilling rig to perform drilling operations as part of its services as a geotechnical engineering firm. In addition, one of the technicians was a Texas-licensed water well driller. Since many of the geotechnical clients had left with the previous Office Manager and employees, I was left with a large void in business. Having this drill rig provided me the opportunity that I needed to rebuild the business for Chenn-Northern. I moved our business operations into the environmental consulting area by doing underground storage tank investigations and after that site remediation services if gasoline leaks were identified. This gave me an advantage over those individuals who had left Chenn-Northern to start their own geotechnical firm since they did not have their own drilling rig.

With the number of Exxon and Chevron gasoline stations in the San Antonio and the surrounding area, it was not hard to get Exxon and Chevron back as clients. The state had established a time limit when all the gasoline stations had to be investigated and remediated if the site was found to have been impacted. Having our own drill rig allowed us to be able to respond a lot faster than the defecting office manager and his employees. They had to use a contract drilling firm to complete the required drilling operations. The local contract drilling firms were quite busy as demand for their services was now great. We could also do the drilling more competitively as far as price. With these advantages, I was able to move the business into the environmental consulting side of operations, which offset the lack of geotechnical engineering projects that were no longer available.

It was not long before our office was doing quite well with our environmental consulting services. We were doing about 80 percent of business in the environmental field and 20 percent in the geotechnical field of operations. I was able to grow the Chenn-Northern office back to the level it was before the desertion of the previous office manager and his employees in about a year and a half. I was learning what the free market capitalistic economic society was all about. I learned that capitalism is a brutal and cutthroat operations environment that exists when conducting business operations. This, of course, was a much different environment than that I had been used to in the military.

The next challenge occurred when Chenn-Northern was bought out by a holding company based in Great Britain. This holding company was using a business model of rapid growth in order to advance their stock prices on the stock market. Their model required a 50 percent growth rate in business operations each year, which resulted in their stock being quite attractive for the stock market. They were achieving this by increasing business operations as much as possible at each office and after that by overall expansion through acquisition of additional companies, like Chenn-Northern. This type of rapid-growth business model can work in the short term but cannot be sustained in the long term.

The difficulty with this type of business model occurs on the expansion side. When additional companies are acquired, the next company that is procured had to be twice as large as the previous one acquired. Chenn-Northern had seven office locations throughout the northwestern portion of the United States when it was acquired by the holding company. Therefore, to continue with this business model, the next acquisition would have to be twice the size of Chenn-Northern.

The holding company's next acquisition was Southwestern Laboratories Inc. (SWL), which had fifteen offices in the southwestern US. It was twice the size of Chenn-Northern and did fit with the holding company's business model.

The challenge for me was that there was a Southwestern Laboratory (SWL) office in San Antonio. This resulted in the Chenn-Northern office being in competition with the Southwestern Laboratories (SWL) office. They had been in San Antonio for a significantly longer period of time than Chenn-Northern. The SWL headquarters was located in Houston, Texas. The San Antonio SWL office concentrated on geotechnical engineering operations and had a solid client base. I was not sure what the holding company had in mind, and I was not sure if I was even going to be able to keep my job.

This is when additional empirical evidence was revealed to me that God had to be guiding my life's path. The decision being considered by the holding company was to combine the two offices under the name Southwestern Laboratories (SWL). The name SWL was kept since the SWL office was about twice the size of the Chenn-Northern office. This meant that there would be no need for two office managers, and the SWL manager would, of course, have priority over me. He had been with SWL for a number of years and was doing a very good job.

The one good thing for me was that the SWL office business operations were primarily geotechnical services, and they also had their own drilling rigs. Nevertheless, they were getting into the environmental consulting side of business operations. Their business operations consisted of 80 percent geotechnical services and 20 percent environmental consulting services. Since the Chenn-Northern office was doing 80 percent environmental and 20 percent geotechnical, the holding company decided to beep both offices open with the SWL office performing geotechnical engineering services and the Chenn-Northern office performing environmental consulting services. We gave up our 20 percent geotechnical clients to the SWL office, and they gave up their 20 percent environmental clients to us at the Chenn-Northern office.

The old SWL office was after that known as the Geotechnical Division Office of SWL, and our office was known as the Environmental Division of SWL. As such, both of us office managers were able to keep our positions as managers of our separate divisions. This had to be God's work in my life and more empirical evidence that God does exist.

The rapid-growth model that the holding company in England was using worked for the next year and a half. The problem with trying to expand business operations at a 50 percent level each year is not possible for individual offices. Expansion by acquisitions, therefore, had to be used to continue the desired growth rate. The problem for the holding company began when it became harder and harder to find larger engineering consulting firms to acquire. This is when the holding company began to require that each location increase their level of business operations by 15 to 20 percent each month. Anyone in the engineering consulting business knows that this level of growth is impossible to achieve. The holding company's main business was in pharmaceuticals, not engineering consulting. This all came to a head for me, three years after I had been hired by Chenn-Northern.

Our normal operations required that each SWL Division office issue a monthly profit and loss (P&L) statement. My boss was located in a SWL office in Austin, Texas. This is when the holding company decided to use "Enron" accounting methods to offset the lack of expansion opportunities through acquisition. My boss called me and directed me to add $25, 000 to my P&L

statement in a category called "work in progress." He said that this would represent work already accomplished on open projects, but the project had not been completed and billed. He said it would represent income that would be realized on our open projects that we had accepted proposals for but had not yet started or completed. He had no idea how many open projects we had nor how many signed proposals we had. The $25,000 number was a number he needed to meet the holding company's requirement with no basis in reality. I told him that I could not and would not do that because this was not considered to be a "generally accepted accounting principle (GAAP)." As a professional engineer with an MBA degree, I knew that this was not ethical and was wrong. He said that I had to do it and hung up on me. The sad thing was my boss was also a professional engineer.

My office accountant normally prepared our P&L statements and after that mailed them to the Austin office. I told her what my boss had requested and that we were not going to comply. I told her to prepare our P&L statement without the $25,000 "work in progress" figure, and I would hand deliver it to the Austin office. I drove to Austin and handed our P&L statement to my boss and told him that I had not included the $25,000 figure because that figure was not real income earned and furthermore may not ever be generated. This, of course, did not go over well with him. I departed the Austin office and returned to San Antonio. It was about an hour-and-fifteen-minute drive back to San Antonio. When I arrived back at the office, my accountant informed me that I had been fired. This was in 1992 and is when God again stepped in and led me in a new direction.

One of the geologists had remained behind with the Chenn-Northern office when the previous office manager and some of the employees had left to start their own consulting firm. He stayed with Chenn-Northern, and he was the individual that had called the headquarters to inform them of what was happening. With my firing, he and I decided to start our own environmental consulting firm. He gave his two-week notice and quit, and we began the process of starting our own firm. God was providing a new path and opportunity for the both of us.

Both my MBA degree and my Texas PE License were of great benefit to us as I prepared our five-year business plan. The state of Texas requires a five-year business plan in order to incorporate a business in Texas. Both the MBA degree and my PE license were opportunities that God had provided me while I was in the military.

When atheists claim that there is no God, I have to feel sorry for them. To me, there is too much empirical evidence in my life that refutes their claims. I have previously discussed how God guided my life in the military, which resulted in a quite successful military career. He also guided my life on the engineering path that he had planned for me during the military and after that my postmilitary life.

God guided me to attend ISU and study aerospace engineering. As part of the aerospace engineering curriculum, we had to choose between structures and propulsion as areas of concentration for study in addition to the required core courses for the degree. There was not enough time to study both of these areas in addition to all of the aerodynamic courses required for the degree. God led me to choose structures, which turned out to be what I needed to become the lead structural engineer for the C-5A fleet while I was in the military. He knew what I was going to need during that phase of my life.

Another excellent example of empirical evidence that God was guiding my life was revealed when he provided me the opportunity to go to AFIT while I was in the military and get a master's degree in engineering. The Air Force never asked me what field of engineering I wanted to study for my master's degree. After Vietnam, when I reported to the University of Denver, I found myself enrolled in the Civil Engineering Department. God had to have been behind this decision since he knew I would need a master's degree in civil engineering during his plan for me in my postmilitary life. He also knew I would need an MBA degree in my postmilitary life, so he guided me to get that degree also.

After I retired and was looking for a job, God led me to Chenn-Northern. The job opening for an office manager at Chenn-Northern had as part of the required qualifications included a master's degree in civil engineering, along with a Texas PE License.

In addition, having an MBA degree was highly desirable, but not required. God knew exactly what I needed in order to be a perfect fit for the position with Chenn-Northern after I retired from the military. To me, the empirical evidence of the above only confirms that God had to be the one making all of this possible as part of his plan for me.

God had prepared me to be able to develop a business plan and be able to start my own engineering consulting firm after I was fired. My partner and I filed all the required paperwork, and STC Environmental Services Inc. was born during the second half of 1992. We rented office space, acquired some office furniture, and began business operations. We began business operations with four employees. Three of the employees were myself, my partner, the geologist, and his wife, who worked as our office receptionist. The fourth employee was the accountant who had worked for me at Chenn-Northern and at SWL.

I had hired this accountant when I was hired to manage the Chenn-Northern office. The previous accountant was one of the employees that left when the previous office manager left to start his own consulting firm. A total of seven people had left Chenn-Northern when the defection occurred.

On my departure from SWL, my accountant that I had hired for Chenn-Northern and now was working for SWL was also fired. She was fired after I was because she complained to SWL management about the improper accounting procedure that she was being directed to do. This was the same procedure that I had refused to comply with. With her hiring as our accountant, we now had our nucleus for the start-up of STC Environmental Services Inc.

I know that God had to be guiding us because it is well known that approximately 75 percent of all small businesses fail within the first three years of operations. It was quite tough getting our business started. My partner and I had both put in funds for start-up costs. With intense marketing efforts, we began to acquire clients. In the environmental consulting business, it takes some time to acquire clients and after that complete projects. We had to prepare proposals, get them accepted, and then complete the project. This all had to be accomplished before the project could be invoiced and before cash could begin to flow.

As our workload began to increase, we needed to hire additional employees in order to be able to complete the work. This is when I know that God was guiding our efforts in our business operations. My partner and I did not receive a paycheck for the first three months that we were in business. It did not make sense for us to pay ourselves out of the start-up funds that we had put into the business. We needed those funds to meet payroll for our employees and to pay operational costs.

The first year of business operations was very—and I mean very—tough. There were a number of times when payroll was due on the morning of payday, and we did not have sufficient funds in our checking account to cover payroll at the start of the day. The night before payday, if this was the case, I would wake up during the night wondering what we were going to do. When this would happen, I could not go back to sleep. I still had not learned that I should not have worried as the Bible states that God knows what we need, and he will provide it even before we pray for it. Each morning on those days when we were short of funds, when I got to work, I would pray for God's assistance. He never once failed me, and we never missed a payroll on payday. Only our accountant, my partner, and I knew how close we came to missing payroll.

At our office location, the mail was always delivered by 10:00 a.m. each day. We normally issued paychecks at the end of the day. On those days when we were short on funds, our accountant would go out to the mailbox hoping that there would be checks in the mail. When the mailman delivered the mail, she would immediately go to the mailbox and after that come back in. If she had a smile on her face, I knew that God had answered my prayer. There were a number of these days during the first year, and he never once let us down. By the end of the first year, we were on relatively firm financial basis for the firm.

I also know that Satan is always hard at work and God tested us. After we started the business and we were working hard to get clients, we received a call from one of the well-known power brokers in San Antonio. He was the owner of one of the four largest

San Antonio shopping malls. He invited us up to his office in his mall to discuss business opportunities. My partner and I both went to see what he had to say. His business would be a large plus if we could secure it.

Once we got to the meeting, it became quite obvious to me that God was allowing Satan to tempt us to see how we were going to respond. This individual wanted us to provide him with environmental reports to satisfy all the new environmental regulations that had been passed. These regulations required a significant number of environmental investigations that would be required for his shopping mall. Through his discussion, it became quite clear that what he truly wanted was for us to essentially "pencil-whip" these investigations in order to satisfy the regulations. As a local power broker in the city, he informed us that if we worked for him, he would ensure that we would receive all of the environmental work from the city that we could handle.

As a PE, I knew that I could not agree to work for this individual. My partner and I declined his offer and departed his office. It was soon after that, that our phone began to ring a lot at the office, and we began to acquire many new clients. God had tested us, and we had passed.

I should note that at the end of our first year of operations, the holding company that had acquired both Chenn-Northern and Southwestern Laboratories went bankrupt. Their stock prices plummeted when they were no longer able to continue with their rapid-growth business model. They had to sell off their companies that they had purchased for ten cents on the dollar.

In addition, after approximately five years, the power broker owner of the shopping mall passed away. His son tried to run the business, but soon failed and had to declare bankruptcy. When a buyer for the mall could not be found, the mall was torn down, and the land sold off the settle the debt that remained. STC Environmental Services Inc. was doing fine. God knew what was happening and took care of us.

I stayed with STC Environmental Services Inc. for 18 years as the President and Principal Engineer. I worked until one day in November 2009, when I woke up thinking I was no longer having fun. I had always been a workaholic, and now I was tired and ready to retire for good.

My wife, Jan, had retired in 2002 from Taft High School in San Antonio, Texas, where she had been the head counselor. She always told me that I would know when it was time to retire. On that day in November, God led me to decide that it was time for me to retire for the second time. I was sixty-seven years old, and I was getting tired. I retired from STC Environmental Services Inc. on February 1, 2010.

I should also note that as of the date of this book, STC Environmental Services Inc. is still in business. That is over 33 years of business operations. In addition to the holding company and all of their acquired companies that were no longer in business, the company that the defecting employees from Chenn-Northern started also went out of business. There is no one that can tell me that there is no God.

# CHAPTER 25

# The Retirement Years

IT IS NOW 2025, and I have been retired for fifteen years. I have never regretted my decision to retire. Jan and I are both enjoying our retirements, and we are very thankful for all the blessings that God has bestowed and continues to bestow on us throughout our entire lifetimes. I know without a doubt that he has guided our lives the entire time and continues to do so even in retirement. This was evidenced in one specific incident and a relationship that we have developed that can only be considered a blessing from God. The specific incident that occurred was when, once again, God saved my life. I used to put out a lot of Christmas decorations in the front yard each Christmas. With each year, I would add more and more decorations to the yard display. The neighborhood Homeowners Association has a Christmas decoration competition contest each year. I finally had so many decorations that our house won the Christmas decoration contest three years in a row. Between Christmases, I would store all of my decorations in our attic. The access to our attic is by a pull-down ladder from an opening in our hallway ceiling. I would have to climb up and down this ladder to get the decorations down each year and after that put them back up after Christmas. The incident occurred during one of these Christmas-decoration events when I was putting the decorations back up into the attic.

I was wrestling with a large heavy container of decorations when I fell backward into the ladder opening while I was working in the attic. I know that it was God who saved me. When I fell backward, one leg was restrained on the opposite side of the opening while my head and shoulders were restrained on the other side. The other leg and the majority of my body were extending through the opening. If I had fallen through this opening, I would have landed on my head on the hallway floor below. I had absolutely no control over the falling motion, and I know it was only with God's helping hand that I did not fall through the opening that day. This incident truly got my attention, and I now store all of my Christmas decorations in the garage. I do not even like going up into that attic anymore.

The relationship that Jan and I developed that we consider a true blessing from God concerns a young couple that moved in across the street from us. This couple had both graduated from Texas A&M University, and they are Aggies. Our oldest son, Doug, also had graduated from Texas A&M and is an Aggie. When the couple moved in, the wife was pregnant with her first child. With them being Aggies and with us being Aggie parents, we had a common bond to start. As they were moving in, Jan and I went over to welcome them to the neighborhood. This is when we found out they were Aggies.

Our oldest son, Doug, and his wife, who both are Aggies, had decided not to have any children. Our youngest son, Randall, is not married and, as such, does not have any children. This, of course, created the situation that Jan and I did not have any grandchildren, and the future did not look like this would change.

In addition to this couple being Aggies, we had a lot in common with them. The husband and I were both engineers. He had worked for NASA before moving to San Antonio. His wife and Jan had both been schoolteachers, which meant they had a lot in common. I know that God had led this couple to our neighborhood as part of his plan for them and us.

The couples' first son was born approximately two months after they moved in. Not having any grandchildren, Jan and I were quite excited and happy to be able to help our new friends with their new son. To us, this young couple became family, and we became quite close to them. I know that this was God's plan and a blessing to us because they have "adopted" us as "grandparents" to their children.

We are now the proud grandparents of their three awesome and wonderful children. I think everyone needs grandchildren during their retirement years in order to have a full and complete life. The sound of their young voices in our home makes our house a home and is a true blessing for Jan and me from God. Playing Santa Claus each Christmas is a true delight for me because of the looks on their young faces.

Retirement has given me the time to reflect back over my entire life. During this reflection, I have been able to truly realize how God has guided me through my life. He guided me to write my first book titled "Hank: An Angel Dog." This book was about Hank, our dog that our family had rescued from an animal shelter. Hank joined our family on the day that my mother-in-law passed away. We got him to help my father-in-law with the loss of his wife. Writing that book helped me realize how God watches over all of his creation. Hank was truly an angel dog.

God also led me to write my second book titled "The Long Return." This book was about my time in Vietnam and the difficulty I had when I returned from Vietnam. We Vietnam veterans were not welcomed home in a very pleasant manner. We were met with a significant amount of antiwar, antimilitary sentiment by the American society. Writing that book helped me understand how our American society had changed, and it helped me find peace within myself. It was quite therapeutic for me. Some of my friends that I gave copies of my book to who had been in Vietnam also said that the book had helped them. This book also allowed me to understand how God was guiding my life as a part of his plan for me while I was in the military.

It took approximately four years of my retirement to write and publish these two books. When I finished those two books, I felt that two books were enough. I never would have thought that I would ever write one book, let alone two. Writing the two books truly got me reflecting back over my life. That is when I truly started to understand how God had been watching over me and guiding me to make the decisions that I did in order to fulfill his plan for me. I was not planning to write any more books until God led me to decide that I needed to write my third book. I figure he has a purpose, and I needed to do it. That's when I wrote "Don't Fly Today."

This was the hardest book for me to write. On my first try, I stopped after about two months. It was not coming together like I wanted it too. It was not flowing, and I became frustrated and stopped. About two months later, I decided to try again with the same results. It was again about two months when I felt that God wanted me to try again. So, I began again and the story flowed and came together like I wanted it too. I think God wanted me to write "Don't Fly Today" for his purpose. It must be part of his plan for someone else whose path that he is guiding.

While I was reflecting back over my life, I began to put the decisions and events that have occurred in my life into perspective. That is when all the empirical evidence that a higher power had to be present and is working for good in this world became undeniable. If a person only looks at each decision or event in isolation, they may not understand how all these events and decisions are actually pieces of a puzzle, which is God's plan. These pieces all fit together to make up his plan for our lives. All we need to do is listen to him and obey.

I now know that my sole purpose and job for the remainder of my life that God has planned for me is to take care of my wife of sixty-three years. She was diagnosed with dementia in 2014 and is in a memory care facility since May 2023. I go to see her every day. I had promised both God and her dad that I would take care of her. God has blessed me, and this is the very least that I can do until he calls me home.